MOTHER COUNTRY

ALSO BY MARILYNNE ROBINSON

Housekeeping (1980)

MOTHER COUNTRY

Marilynne Robinson

Farrar, Straus & Giroux

NEW YORK

Library of Congress Cataloging-in-Publication Data
Robinson, Marilynne.
    Mother country.
    Bibliography: p.
    1. Reactor fuel reprocessing—Environmental aspects—
Great Britain.   2. Nuclear power plants—Environmental
aspects—Great Britain.   3. Radioactive pollution—
Environmental aspects—Great Britain.   I. Title.
TK9360.R65   1989   363.1′79   88–33523

For Fred, James, and Joseph
With thanks to Melissa Gordon

While nations consent to put into any hands an uncontrollable power of mischief, they may expect to be thus served.

Jeremy Bentham

MOTHER COUNTRY

Introduction

Perhaps the real subject of this book is the fact that the largest commercial producer of plutonium in the world, and the largest source, by far, of radioactive contamination of the world's environment, is Great Britain—and that Americans know virtually nothing about a phenomenon that occurs, culturally speaking, so very close at hand. The primary producer of plutonium and pollution is a complex called Sellafield, on the Irish Sea in Cumbria, not far from William and Dorothy Wordsworth's Dove Cottage. The variety of sheep raised in that picturesque region still reflects the preference of Beatrix Potter, miniaturist of a sweetly domesticated rural landscape. The lambs born in Cumbria are radioactive. This fact is ascribed to the effects of the Russian nuclear accident at Chernobyl, but Sellafield is so productive of contamination that there is no reason to look elsewhere for a source. Testing of lamb and mutton was only undertaken some months after Chernobyl, though the plant at Sellafield routinely releases plutonium, ruthenium, americium, cesium 137, radioactive iodine, and other toxins into the environment as part of its daily functioning. The fact that food had not been tested systematically in an area whose economy is based on the production of food as well as the production of plutonium is characteristic of British policy, wherever there is a potential impact of industrial practice on public health.

It should be noted that the plant at Sellafield was built by the British government. It was developed and operated by the U.K. Atomic Energy Authority, and then given over to British Nuclear Fuels Limited, a company wholly owned by the British government. It should be borne in mind that the plant receives waste and reprocesses plutonium for profit, to earn foreign money. Sellafield is at the center of an economic configuration of a kind as yet unfamiliar to Americans. It is a part of the electrical-generating industry because it absorbs the wastes produced in British reactors, transforming them, in part, into salable materials through reprocessing. It expedites the sale of British nuclear technology abroad by accepting wastes generated in other countries, the costs of engineering services and waste disposal lowered by the value of these reprocessed materials. It is a closed cycle (putting aside the fact that the public subsidizes it once in the price they pay for electricity and again because of its military role as supplier of plutonium for British bombs) in which each stage stimulates profitability in the others. To call a government-run industry highly profitable, when on the one hand it is the monopoly supplier of a very costly product, as electricity is in Britain, and on the other hand it is incalculably destructive of the public health, seems to cause no embarrassment to the plant's defenders. The British nuclear industry creates leukemia in the young and hypothermia in the old, and yet it is profitable. Clearly bookkeeping is as expressive of cultural values as any other science.

Sellafield has flourished in the care of Labour and Conservative governments alike for thirty-five years, during which time it has poured radioactive wastes into the sea through a pipeline specially constructed for that purpose, creating an underwater "lake" of wastes, including, according to the British government, one quarter ton of plutonium, which returns to shore in windborne spray and spume, and in the tides, and in fish and seaweed and flotsam, and which concentrates in inlets and estuaries.

The plant is expanding. Wastes from European countries, notably West Germany, and from Japan, are accumulating there, while the British develop means of accommodating the pressing world need for nuclear waste disposal. Their solution to the problem amounts to extracting as much usable plutonium and uranium from the waste as they find practicable and flushing the rest into the sea or venting it through smokestacks into the air. There are waste silos, some of which leak uncontrollably. In an area called Driggs, near Sellafield, wastes are buried in shallow earth trenches. Until the practice was supposedly ended in 1983 by the refusal of the National Union of Seamen to man the ships, barrels of nuclear waste were dropped routinely into the Atlantic. In other words, Britain has not solved the problem of nuclear waste, has in fact greatly compounded it, in the course of producing plutonium in undivulged quantities.

What happens to this plutonium once it is extracted no one says. We must depend on the wisdom and restraint of the British government to keep it from falling into the wrong hands. Yet one arrives fairly promptly at the realization that the only prospect as alarming as that all this plutonium should fall into the hands of irresponsible or malicious people is that it should remain in British hands. The essential act of irresponsibility is, after all, to have produced it in the first place. And then the British are not especially fortunate. Sellafield itself has had about three hundred accidents, including a core fire in 1957 which was, before Chernobyl, the most serious accident to occur in a nuclear reactor. Sellafield was called Windscale originally, until so much notoriety attached itself to that name that it had to be jettisoned. That an accident-prone complex like this one should be the storage site for plutonium in quantity is blankly alarming.

There are many useful lessons to be learned about the nature of contemporary history from the study of Sellafield. Most informed Americans believe that the release of plutonium on an important scale into the environment would entail disaster of world historical proportions. Yet every account of our present situation sees grand-scale plutonium contamination as a threatened consequence of the competition of the so-called superpowers. In other words, the account we make of present history is radically in error, not least in the matter of the importance of the United States and the Soviet Union in determining the fate of the earth. If plutonium deserves its reputation, then a nuclear war will simply accelerate the inevitable. Statesmanship in Moscow and Washington will merely postpone the inevitable. This is to say that decisions of the greatest consequence have been taken, while our savants and moralists looked resolutely in the wrong direction. We have pondered the Russian soul, and our own, and we have seen the darkness in them both as deviation from the human norm. Western Europe and especially Britain we have assumed to be mild with age, peripheralized by the drift of history. Yet any final reckoning would probably find Britain's impact on the postwar world greatest of all nations. Fleets sail away, ideologies talk themselves to death, empires yield to cultural tectonics. But plutonium is everlasting, for human purposes, and it is irretrievably a part of the world environment, because the British have made a business of pumping it into a shallow sea through a pipeline a mile and a half long, and have prevented no one from fishing in the area, though the fish are radioactive, five thousand times more so than fish caught in the North Sea, though that is also contaminated. They have prevented no one from living or vacationing there, or growing and marketing food in a countryside affected by radioactive wind and rain. Plutonium from the plant carried by the sea has already been found in Ireland, Iceland, Sweden, Denmark, and Belgium.

I am aware that the situation at Sellafield raises a great many questions as to how and why such a thing should have come about. But the fact of the plant's existence and operation is not disputed—It can be confirmed by anyone in America who cares to spend a few hours in a fairly good library looking at British publications. The mystery is not how a phenomenon of this importance can be concealed but how, being, as it were, a city on a hill, it has remained unknown to us for so many years.

There are certain questions I have not attempted to answer. The first of these concerns the role of the American government in this enterprise. If it has no role, then it is virtually alone among the major Western governments, most of whom use the plant's services. There is mention in some British sources of a barter arrangement in which between 1964 and 1971 the United States received Sellafield plutonium in exchange for nuclear materials produced in America. This in itself means little. The fact that Britain produces plutonium as part of its role as a nuclear weapons state is no secret, nor is it ever surprising to find ourselves and the British tinkering at these projects together. The question is whether the American government has encouraged the operation of Sellafield, an unaccountably foul enterprise, even by the debased standards by which such things are judged. I am troubled by the thought that the United States must have a satellite or two sensitive to concentrations of radioactivity, and that the contamination of the world's hottest sea cannot have gone unremarked. The Soviet Union must also have means to detect radioactivity in significant quantity. And presumably it would be to their advantage in the battle for hearts and minds to point out to the people of Europe how they are being poisoned, especially if there is a significant American involvement. The silence in which Sellafield prospers is a little uncanny. It suggests interests are being served that are neither ideological nor national.

The most plausible hypothesis, I suppose, is that individuals who do not feel any kind of loyalty to the future have been corrupted by the quantities of money involved. This is only speculation. But the currency that passes in nuclear transactions tends to be denominated in millions and billions, and the industry worldwide is protected by secrecy and by its significance in maintaining the prestige of governments and by its military significance, whether as licit or illicit supplier of fissile materials or as potential target. Only assume the usual human slovenliness and venality and an important degree of corruption will seem quite probable.

The nuclear industry enjoys the respect generally

accorded to science, because its workings are abstruse. Everyone knows that it is impossible to predict or to describe the impact of this industry on the planet, that it is persisted in despite its demonstrated potential for disaster, and that, if it were closed down tomorrow, even assuming everything goes well for the next score thousand years in controlling the virulent materials it will leave behind, its economic cost to future generations will utterly dwarf whatever value it has had for our own. Yet despite all this the nuclear enterprise is accorded sufficient respect to make the suggestion that its development might to some degree reflect ordinary corruption seem a little impolite. It is as if money and secrecy brought out the best in human nature, changing politicians and technicians into Carmelites.

But Sellafield, if it existed in isolation, would be sufficient proof that the world's interests have not been properly respected. As evidence of culpability, it is not a smoking gun so much as a hail of bullets.

Objections are made to Sellafield's operation by governments within Europe. Denmark and Ireland, whose small terrains are increasingly affected, protest vehemently in the European Parliament. Other countries join with them. But this is disingenuous, since these other countries—unlike Ireland and Denmark, who have no nuclear power plants—patronize Sellafield. They pay Britain to take on their waste disposal problems. Their indignation tends to obscure the simple fact that European wastes are poured into the European environment, day after day, as the result of those same methods they deplore, and that the operation of the plant is profitable because they pay Britain to transform wastes from these countries into reprocessed plutonium, and uranium, and all the varieties of contamination for which Sellafield is renowned. And they pay Britain to store wastes while new reprocessing facilities are built at the site, or until sea dumping, if it has been stopped, is resumed. Circumstances demonstrate that it is politically possible in Britain to do these things, flagrantly damaging as they are to the national and the world environment. Circumstances prove also that the superior environmental standards of other countries are shadow play, since toxins are turned over to Britain to be dispensed with by methods these same countries deplore so loudly, the consequences of which are already being felt by their own populations. The efficacy of respective European environmental movements can be established from the fact that Sellafield is in Britain, and that German, Italian, and Scandinavian wastes are at Sellafield and in the waters off the European coast.

There is an analogue for Sellafield in the disposal industry that has developed in Britain for non-nuclear toxic wastes, which are shipped into the country and then disposed of in the variety of ways the laxity of law makes possible—left in municipal dumps or poured into the North Sea or buried on derelict land in poverty-stricken northern regions—40,000 tons of lethal waste being imported in 1986.* What little fastidiousness is reflected in other countries' hiring these services merely makes it profitable to the British government to leave its environment undefended. The arrangement creates an incentive for employing the crudest means of disposal— for which Europe and the world will suffer in due course.

All this raises the question of the role of environmental groups. People in this country have felt for some

* See "Unions try to stem rising tide of toxins," *The Observer*, August 2, 1987, p. 5.

time that Greenpeace, Friends of the Earth, and the Sierra Club, as well as scientists' organizations, were keeping watch over the greater environmental issues, at least to the extent of making them known to us. Sellafield is the world's largest source of radioactive contamination. It is also the greatest commercial producer of plutonium in the world—a role disturbingly at odds with any notion of non-proliferation. It is the center of a worldwide traffic in toxic and explosive materials. And it is the site of the largest construction project in Europe, because it is expanding. The release of radiation from Three Mile Island is usually estimated at between fifteen and twenty-five curies of radioactive iodine. Many hundreds of thousands of curies of radioactivity have entered the environment each year through Sellafield's pipeline and its stacks, in the course of the plant's routine functioning. In other words, Three Mile Island was a modest event by the standards of Sellafield and, even if disaster had not been averted, would have made no comparable impact on the earth's habitability.

Yet, though Greenpeace is deeply involved with Sellafield in Britain, even to the extent of having supposedly placed a mole inside the plant, information about it in the United States is extremely limited and of poor quality. The plant is located in England's largest national park, which includes the Lake District. Tourism is an enormous industry there. It seems to me indecent that people are not warned away from this uniquely contaminated environment.

The British government cannot be expected to show foreigners a solicitude it does not show its own people. But the American government cannot be excused for allowing its nationals to be unknowingly exposed to pervasive contamination—for example, to breathe air in which concentrations of plutonium are sometimes higher than they are within this old, disreputable plutonium factory itself.

Greenpeace, since it enjoys the goodwill and the financial support of a great many people in this country, surely owes them a reasonably accurate description of the state of the world, as well as information which could directly affect their own and their families' well-being. If Greenpeace takes exception to the usual accounts of the lethal properties of plutonium—for example, that a particle invisible to the naked eye, if inhaled into the lung, will ultimately cause a cancer—then they should tell us so. If they adhere to the common view, then they should explain why they allow tourists to wander into an area where this misfortune is so liable to befall them.

It seems to be the policy of Greenpeace to compartmentalize its activities, at least to the extent of keeping Americans preoccupied with issues that arise within our borders—and with aquatic fauna, whose trials and troubles seem never to include ingestion of radioactive fish in the Irish Sea or the North Sea.

If the decision to publicize environmental problems selectively is tactical, then it is past time for Greenpeace to admit that the approach has failed. The greatest source of radioactive pollution in the world is growing exuberantly, fed by an influx of yen that makes it Britain's greatest earner of that potent currency.

It is in fact difficult to imagine any strategy that could have produced a less desirable result. While the effect of relatively stringent environmental laws in Europe has been to siphon waste into Britain, Greenpeace chalks up triumphs of environmental consciousness without reference to the fact that the scene of contamination has merely shifted, not very far, and that a powerful economic reward has been provided for the recalcitrance of the British government in environmental matters. These tactics make it difficult for Sellafield's foreign clients to use crude disposal methods at home, but do not put pressure on them to develop responsible methods, or indeed to accept responsibility for the environmental consequences of their own industries. Britain can bear the opprobrium of Europe and Europe can endure the malfeasances of Britain. The nuclear industry can enjoy the cheapest possible solution to the problems of waste—the costs of dumping are, you will recall, indemnified by the production of uranium and plutonium, the latter a valuable commodity for reasons that are never quite made clear.

If information published by Greenpeace in Britain were published by Greenpeace in America, the system might not work so smoothly. It would be highly sensitive to reaction in this country manifested in reduced investment and tourism, a reduced enthusiasm for European products, an increased skepticism as to the disinterested wisdom of European governments, and perhaps a less reflexive confidence in the community of values which is always adduced to persuade us that in defending Europe—and especially Britain—we are only defending ourselves.

If it is a tactic or strategy to select and ration the truth, in order to direct public reaction toward ends the organization considers desirable, then they have violated the most basic tenets of democracy, even while producing a spectacular vindication of the wisdom of these tenets. Sellafield is a disaster. It violates common decency and common sense. It is the sort of thing that withers under informed public scrutiny—from which, to the misfortune of this beleaguered planet, it has been sheltered in America. The regularity with which foolish and destructive policies are concealed from the public is the most powerful statement possible of what democracy could have meant to the world, if there had been the courage and patience to sustain it.

The considerable involvement of Greenpeace and Friends of the Earth, which is associated with the Sierra Club, should mean a flow of exact and urgent information about Sellafield into the American environmental movement and its publications. Something else has happened. In 1984, the Sierra Club published a book by Walter Patterson, called *The Plutonium Business,* which drops its tone of objective description of nuclear facilities in other parts of the world to attack a headline in the London *Daily Mirror* (October 1975) which said Britain was to become the world's "Nuclear Dustbin." This characterization is in fact commonplace and fairly precise, considering the service Sellafield performs for Britain and the world at large. Reading *The Plutonium Business,* one would never guess that *Britain* is the center of the world plutonium business. Walter Patterson is active as a writer on nuclear questions in the British press. His omissions cannot reflect ignorance. The Sierra Club should also be aware of Sellafield in some detail, if they are in communication with Friends of the Earth in Britain, since this has been an issue of overwhelming importance in Britain and Europe for a number of years, highlighted by the leukemia deaths of children on the English and Irish coasts. Friends of the Earth figures in many inquiries and demonstrations. While it is possible to imagine that these organizations do not communicate with their associates, or pool information, and that they are by intention parochial and intramural, it behooves them to make this clear, and to abstain from publishing books like Mr. Patterson's which mislead by seeming to deal in an authoritative way with international phenomena.

The silence of the American press (there was an excellent article in *Newsday* by Patrick J. Sloyan, May 20, 1986) is consistent with a more general failure to report news of substance from abroad. Nothing is stranger than to live in Britain and read what American newspapers and magazines print about it. If it were the sworn duty of the American press to render the United States incompetent in every aspect of foreign affairs, our journalism would be very little different from what we have at present. The hundreds of reports about Sellafield in major British newspapers and on television yielded slight, late, perfunctory articles by Joseph Lelyveld in *The New York Times* and Karen DeYoung in *The Washington Post,* both of which concluded that it was all a tempest in a teapot, more or less. The locals (those people whose children tend to die in disturbing num bers) felt no concern about the plant, according to these articles. DeYoung notes that the many tourists "never seem to mind the site." *Times* and *Post* readers no doubt. The anxiety, these articles agree, is all in London.

People from the area have in fact flung Cumbrian silt through the door of 10 Downing Street, and seen it cleaned away by men in radiation gear. People from the area have raised money to buy Geiger counters, so that they will not have to depend on the government for information about their circumstances. There have been demonstrations, strikes, and votes of no confidence in the management at intervals over thirty years.

Neither the *Times* nor the *Post* mentions the lake of

plutonium and other radioactive substances which lies off the Cumbrian coast. Both mention alarm caused by leukemia cases too few to be statistically reliable, repeating a bitterly disputed claim by the British government; that is, by the plant's owners and operators. In fact, one child in sixty dies of cancer in the village nearest the plant, and rates in other villages in the region are comparable. To find ambiguous these high rates of a radiation-induced illness in a radioactive environment seems to me willful at best. But the American press has a tourist-bus mentality, a keen and persisting interest in pubs and arts festivals, and will seek out what it considers "typical" and "authentic" while one example of either remains on earth, at the same time ignoring whatever fails to confirm its very banal expectations. There is nothing sinister in any single instance of failure in an institution which fails routinely.

The problem of how an enterprise can prosper, pouring plutonium into the world environment, when contamination of this kind is held to be among the worst potential consequences of a nuclear war, perplexes me deeply. I experience resistance from my hearers, as often as I raise the subject of Sellafield, even though I can document what I say from the most reputable British sources. An American marine biologist assures me that Britain has more naturalists per capita than any other country in the world. I am sure he is correct. This does not mean the British landscape or population enjoys an abundance of informed solicitude. It means that two apparently incompatible tendencies exist side by side. The character of Sellafield is not in dispute. And since Britain is distinguished among European nations for the degree of its contamination, both radioactive and chemical, it is perhaps appropriate to ponder the attributes of British naturalist enthusiams which permit, and disguise, an unaccountably brutal indifference to nature. This nation of birdwatchers and dahlia fanciers uses 2,4,5T, the dioxin-contaminated defoliant banned in every other Western country because dioxin is, like plutonium, often called the most potent manmade toxin. Why should Britain go to the lengths it does to keep rabies out of the country, bothering over lap dogs, while importing chemical wastes? To the extent that dramatizing one highly controllable problem creates an impression of caution and fastidiousness in matters of public health, the illusion is dangerous. For just the same reason, those famous naturalists constitute the opposite of a defense of nature. Britain is a country where information about industrial pollution in drinking water is kept from the public on the grounds that divulging its composition might reveal a trade secret important to the polluting firm's competitive success.

The reader will object that such a policy is not consistent with any reasonable conception of public interest, or even of profit. However, it is typical and pervasive. Beside Sellafield itself, I can adduce the intractable torpor of the British economy as evidence that Britain pursues a miscalculated industrial policy. Since it is no departure from the norms of Britain's endless industrial past, there is every reason to look to its progressive effects on the public health and spirit for a cause of low worker productivity, before blaming excessive security supposedly induced by Britain's so-called socialism, as it is conventional to do.

How and why do the British people accept this monstrosity in their midst? It is they who bear the brunt of it, after all. And they are very fluent in the language of decency, and should notice these trespasses against the sum of things to be valued in this world, from organic life to the health of children to the survival and genetic integrity of species, which include themselves and their descendants.

I have before me a "personal message" from HRH the Duke of Edinburgh, on behalf of the World Wildlife Fund, an organization presently harvesting the largesse of our wealthy country to support exotica abroad. It is part of an advertisement from *Time* magazine, more than half taken up by a bust of the Duke, looking the spirit of patrician high-mindedness. The text is an appeal for support in protecting the planet, on the grounds that "all life on earth is inter-connected, dependent upon the physical processes taking place in the atmosphere and the oceans," and that "if we damage any part of it we are putting our own survival at risk." The Duke's reasoning is as impeccable as the knot of his tie. But he carries understatement to a wild extreme when he adds, "We need people in positions of political power to take into account the needs of nature in their decisions." If His Highness were to point out to Her Majesty that Her Majesty's plutonium factory and radioactive waste dump is the biosphere's single greatest affliction, then she might be able to pull a few strings, . say a word. Who knows? A modest beginning could be made. Even a quiet elderly couple can contribute something. The Queen might refrain a little from knighting people who have distinguished themselves in the waste-dumping line, for example.

But for the British, charity never does begin at home. The logo of the World Wildlife Fund is a panda. Beguiling, and remote, and somebody else's business. It would be more to the point if the logo were a beguiling Cumbrian child. They are the ones in need of looking after.

This advertisement is an instance of the image of reasonableness English people project very successfully. I believe they induce in themselves an enormous moral security, which always prevents them from faulting themselves for anything worse than stodginess or ineptitude or excessive vulnerability to foreign influences. Hearing themselves expound as slick as you please on every great question of the age, abhorring racism, despising the thought of nuclear deterrence, scorning nationalism and militarism, appalled at the spectacle of poverty, they must feel that their gift to the world of moral enlightenment exculpates the racism, poverty, nationalism, and so on with which their own country is grievously afflicted.

People are always inclined to accept an idealized version of their country as its soul and essence, and in cases where this encourages a correcting of institutions toward a higher standard the impulse is valuable. But it is clear that something else has happened in the case of England, because articulated values are so often diametrically at odds with practice. As in the matter of the huge population of naturalists in the most abused landscape in the industrial world, the appearance of enlightenment and benign engagement is the protective coloration for behavior that is marked in an extraordinary degree by the absence of both.

It is not fair to blame the British for deception, when our own press and environmentalists have simply failed to make use of readily available information. Neither does it seem fair to speak of British hypocrisy, since the virtuous utterances with which they chasten and adjure us all are the speech of a people who believe most sincerely in their own decency, from which they do not seem to feel they have departed when they pour a little more plutonium into the environment. They interpret their problems always as arising from an excess of virtue. They are too mild, too courteous, too cautious, too respectful of tradition, too imbued with non-materialist values, too offended by the spirit of competition. Burdened as they are by all this decency, they must cut a few corners to hold their own against other nations not similarly handicapped. In 1976, after extensive discussion of the threat commercial plutonium production would pose to the environment, national security, global stability, and democratic institutions, the *Times* observed editorially that if a plutonium economy developed, no country would be immune to its effects. In other words, the dangers of the trade were too great to act as a deterrent, since for Britain to abstain from it would supposedly not reduce them.* This is typical. Imagining the worst of others, Britain can be the first into the field in the darkest endeavor, confident that the harm it does is excused by being inevitable in any case.

Poisoning the well might seem a radical step, but it is really an established practice never departed from, a bit of nineteenth-century tradition. Industrial poisoning is much older than any concern about it, after all. And the Victorian period, the height against which the British measure themselves and their decline, was a period of utter rapacity. It was during those same brutal decades that the arts of moral refinement were brought to an exquisite polish in the drawing room and the novel. People now regard those days with nostalgia of an especially urgent kind, Americans as well as British, and they forgive eagerly all the wretchedness that consumed the lives of most people, in admiration of the sonorities,

---

* "The Plutonium Problems," editorial, *The* (London) *Times*, October 28, 1976, p. 17.

the niceties, the modest elegances developed over against the misery out of which these rarefied experiences were created. It sounds churlish to point out that Jane Austen's landscape teemed with starveling agricultural laborers, whose misery seems to have been complete, just as it does to note the discrepancies between the wise and gentle urgings of the Duke of Edinburgh and the profit-motivated violence toward the earth committed by his own government. There is a prettiness that takes precedence over reality, that commands a higher loyalty, that readily takes on attributes of moral normativeness even while the conditions of its existence are peculiar and exclusive, violent and corrupt.

If the spectacle of high-mindedness did in fact create norms and refine standards of behavior, Britain, so replete with high-mindedness, would not be profiting from a special willingness to dump toxins over its landscape. Presumably Japan ships its detritus from the other side of the world because no one nearer home can be found who is willing to accommodate it. So Britain's position in the world is apparently rather singular. Its exceptional moral facility actually corresponds to an exceptional willingness to expose its population to harm and hazard others will not bear. (The British government would argue that these wastes pose little risk—that is inevitably the defense—but then the preference of sophisticated nations for dumping them abroad is hard to account for, as is their willingness to pay generously to be rid of them, the source of the very profitability by which, in the mind of the British government, this industry is justified.)

*   *   *

I suppose our situation in America is essentially colonial. As colonists we were the groundlings of other societies, and we are still overawed by the squire, gawkishly eager for a nod or a word. At one time enthusiasm for the common man seemed to be abroad in this land, although we have never been the democrats we claim to be. Now, increasingly, ordinary people, those who are not educated or highly skilled or affluent, are represented as a great reservoir of pathology, crudeness, belligerency, vice, and malice. Everything disturbing in modern culture has been ascribed by the people who deal in such great questions to this hopelessly corrupted mass—one of the great events in the history of intellectual dishonesty.

The cultural origins of our problems are not to be found in the folkways of the powerless, dire as these are at worst. They are not to be found in criminality of the kind that makes us fear dark streets. Young men who yield to the furies and are hostile and violent are no problem beside the phalanxes of diligent operatives too well paid and respected to imagine themselves capable of any antisocial act.

It is clear that American scientists and journalists are aware of Sellafield. Academic specialists from this country have testified at inquiries in Britain into the effects of radiation from the plant. These experts reliably assure the British that radiation is deleterious, and afterward seem to feel that their obligation has been discharged. But the tacit connivance of their silence is simply typical of the response of the thousands of Americans who must by this time know about the plutonium dump off England's coast.

So very much misfeasance is not compatible with the idea of actual conspiracy. I incline to ascribe it to a flaw in our national character. Americans abroad hope so wistfully for approval that they are in effect seduced by the least acceptance, and dashed by the slightest rejection, a weightless people incapable of seeing and judging, as if stuck forever in the most desolate straits of adolescence, merely wishing to be liked and accepted, considering the world well lost if, before the lights go out, they can have a murmur of approval from some foreign person.

This nullity is more contemptible than honest crime, not only because it is a greater falling off from standards of personal dignity, but also because its consequences are more disastrous by any mode of reckoning. The young men who crack heads and strip cars in our dark streets make no claim for themselves as moralists. Atlanticists, on the other hand, make very great claims in that line, taking themselves to be enlightened precisely by exposure to a gentler, worldlier, and less materialist value system. They presume to pity those bad young men who are not, like themselves, refined by experience and civilized by education. They fret because at random babies are fathered and neglected and become in their turn bad young men. They do not fret that babies are poisoned in the womb. That is the work of fine old men, in a land of naturalists and dahlia fanciers—gentlemen who never raise their voices, and whose dress is as reserved as their manner. To call what such men do violent, or corrupt, or degraded, involves a great wrench. The difficulty is merely a measure of our error in fixing our fears on the crimes of the powerless, while grave public men poison the cup we all must drink from sooner or later. In America, we consider it a crime to contaminate the environment for profit. In Britain, profit is considered a public benefit that justifies any means by which it may be realized, every industry being defensible in the degree that it is profitable. Americans in Britain apply British standards, without reluctance and without cynicism, at the same time heartily glad to have shaken the dust of capitalism from their feet, to have come to a place finally where profit is no god, to a non-violent society, a community of goodwill and mutual obligation. For this is, despite all, how Americans persist in viewing England.

The same *Observer* article that describes the flood of toxic chemical wastes from the rest of Europe into Britain ascribes to the government two defenses of this flourishing industry. The first, inevitably, is that to curtail it would cause unemployment and economic dislocation. There is a company in Liverpool with ships built for the job of dumping toxins into the North Sea—a British specialty. To disrupt this enterprise would be a great loss.

The second defense is that if waste dumping were banned, the dumping would only be done secretly, depriving the government of tax revenues, I suppose, since legal and illegal dumping of toxins can hardly differ greatly in any other way. If organized crime were to take on this profitable service to the industrial economies of Europe, what would it do differently? The government is in effect giving notice that it will not effectively enforce any ban on the dumping of toxins— reasonably enough, since it is willing to tolerate and defend the practice. In other words, government policy will prevail through act or omission. A laissez-faire government can practice tyranny by default. This theme will recur.

To appreciate the elegance of all this, the reader must be aware that in Britain there is still no legal obligation to inform a future developer that a site has been used as a waste dump.* So if the bleak cities of the North enjoy some reversal of their economic situation, they may look forward to excavations into pockets of undescribed contamination. The dumping industry adopts the genial theory that a maximum of dispersion through the environment is to be hoped for, reasoning that dilution will attenuate the toxicity of these poisons. This practice has been a boon to the bottled drinking water industry.

Next to the article about importing toxins is another, beneath a picture of a swan on the river Tees.† Somewhere behind the swan are two enormous open concrete vats. These vats contain toxic chemicals contaminated with uranium, and they are "weeping." No one can decide what to do with them. The German owners proposed emptying them gradually into the river, another instance of the notion that toxins are controlled by dispersion. This plan met local resistance. Between the wind and the weeping, however, the dispersion of toxins into the environment will surely go forward on its own. All these strategies of dilution, undertaken in one little island which has gallantly assumed the burden of the most noxious wastes of huge industrial economies, cannot finally be considered dilution at all but, instead, commingling and compounding.

The British government must know why other countries will pay good money to be rid of the wastes Britain imports. So the question arises—do they imagine no future for their own country? Is it simply to be exploited to death? Britain is the greatest source of acid rain in

---

* See "Poison dump to be sold for housing," *The Observer*, November 22, 1987, p. 3.

† "Sweeping threat to Seal Sands swans," Victor Smart, *The Observer*, August 2, 1987, p. 5.

Europe. The government does not put filters on its coal-fired electrical plants because no economic use has been found for the filth they would trap, which it would then be an expense to dispose of. So forests die, and stone decays, and the damage to the landscape and public health entails little expense in money because little is done to defend against such damage. Would these policies be acceptable to any government that looked forward even twenty years?

I know I will shock my readers with my speculations. But more is at stake than their goodwill, though I value approval as heartily as anyone. The fact is that the role the British government has imposed on the country is not finally compatible with its survival. The functions the other Europeans have allowed it to assume are not compatible with their own survival, ultimately. And in the long term, the health of the planet is severely compromised, at very best.

Is there any reason to believe the British are entirely exceptional in adopting such strategies of self-destruction? None that I know of. It seems probable to me that other methods, just as outlandish and unthinkable, for scraping together a few billion dollars in contempt of everything we take to be of value are flourishing in other regions, eluding our notice because the world's calamitous history has not alerted us yet to the profoundly destructive tendencies of this sad species. We are still naïve enough to believe that there is a difference between peace and war, in terms of their impact on the environment and on human prospects for survival. Clearly, if the world is to be preserved, our thinking must change altogether, to take account of phenomena which lack even the abysmal logic of war. We must look at ourselves, and at those we trust and admire, assuming nothing on the basis of such notions as "Western," or "advanced," or, for that matter, "Third World" or "socialist." There are many people of goodwill ready to take action against destructive forces where they perceive them. But Sellafield, the plutonium factory, which gathers radioactive material from the ends of the earth and then pours it into the environment of Europe, is a thoroughly sufficient proof that the world's would-be defenders neither see nor perceive nor hear nor understand. The most important and difficult thing to believe, out of all I will say, is that Sellafield is no secret. Most of what I write about comes from newspapers, often from the front pages of newspapers. It is said that American tourists buy 30 percent of the books sold in Britain. This indicates the kind of travelers who go there—people literate in English, and inclined to read. Is it possible to conclude otherwise than that our education produces an acculturated blindness which precludes our taking in available, unambiguous information if it is contrary to our assumptions?

This book deals with Sellafield and the peculiarities of British culture which allow it to flourish. It deals also with British and American social history. The nuclear threat to civilization and life is a subject talked almost to exhaustion. We think we know something about our "Anglo-Saxon heritage," which we take to be loosely comprehended in the more amiable features of our own society, never implicated in its darker side. Marxism, which I will touch upon, has the chic of a modern, brave, and dangerous philosophy, but Marx is unread, and the versions of his thinking with which we are wearied are the opportunistic inventions of the sort of persons who love to believe they are brave and dangerous. Marx is by no means the source of the vocabulary we routinely use to describe the world. His reputation for intellectual giantism has been put to the uses of notions that were never his. So a slovenly, dishonest, self-congratulatory enthusiasm has affixed itself to him parasitically. His totemization is primitive nonsense, a major example of the necrosis in American intellectual life.

Looking at Sellafield and modern Britain, at British social history, and at Marx, I hope to disrupt assumptions which are important in America, and which are without basis. The subjects I have chosen do not together make a shapely book, but I forgive myself these formal inadequacies on the grounds that very large questions have had to be dealt with in order to anchor the fact of Sellafield's existence in the reader's understanding. I know that Sellafield will be dismissed, if it can be, on the grounds that Britain is a mild and decent society, and that while the plant developed and assumed its economic role, Britain claimed to be a socialist society. Oddly, these notions are potent enough in the minds of many people to mitigate the offense, as if ferocious plutonium, when it is the off-scourings of a government-owned factory pouring into the environment of a virtuous and public-spirited nation, takes on the character of its surroundings and becomes rose-hip tea.

I will suggest that the matter be looked at from another side, that the plutonium should be seen to cast doubt on the benevolence of the British government, and more generally on the legitimacy of the notion that government-run industries are less grasping than others, or that modern governments are reliably more benign than their nightmare progenitors in other ages. In Britain Sellafield's profitability is considered a sufficient answer to every objection. The prospect of diminished profitability is sufficient to postpone improvements in the functioning of this old and primitive plant. The general, long-term effects of government policies, including economic stagnation, meager social benefits, and high unemployment, give the government great leverage, because people are desperate to work on any terms. As industrial employer, the government reaps rewards from its failures, if they are indeed failures, and not simply strategies for creating a cheap and docile work force. Could this be the key to interpreting the sorry condition of the working classes in other countries where the government is also the industrialist? The stability of these countries does not suggest failure, from the point of view of the existing regime. Why do we persist in assuming that any government has the welfare of the mass of its people as an object, where neither history nor present experience encourages this idea?

Can Sellafield exist in a country that is truly committed to economic justice? The plant's great and often-invoked contribution to the economy of depressed Cumbria translates into exploitation of an area especially vulnerable to such abuse, just as it would if the same "contribution" were made to an impoverished region of Africa or Asia. Is a government that knowingly exposes its people to radioactive contamination for profit an appropriate provider of health care or reporter of health data?

Other questions come to mind. What, at this point, can possibly be meant by the defense of Europe? Have we bent our efforts to ensuring them the leisure and freedom to destroy themselves? Will anyone defend the notion that nuclear stalemate has brought us forty years of peace? Or must we give new meaning to the phrase "cold war," seeing that the desolations of war will have been achieved in an atmosphere of quiet and civil order, by the cool, unforced policies of stable and prosperous governments?

It is instructive to consider the content, in practical terms, of the economic statistical categories that are used to compare the successes of governments, and even, bizarrely, the moral solvency of peoples. To the extent that they are accurate, such statistics do not necessarily reflect economic activity as we usually think of it, barter in soybeans and clock radios. Any fluctuation upward in British employment reflects in part the growth of the toxic waste-dumping industry. Any positive movement in Britain's balance of trade reflects in part its successful marketing abroad of these wretched services. Japanese employment and balance of trade must then be thought to suffer a corresponding setback. Yet a sane method of accounting would express all this as aching deficit on the side of Britain.

It is time we developed ways of describing the world which can give us a better sense of its health and prospects. I am not the first to observe that there is, so far as we can know, only one living planet. And even if there were another, nothing in our present state of consciousness would save it from the abuse that threatens to kill this one.

My history of modern England is based largely on newspaper reports, usually contemporary with whatever is being described. Since the British impound all government records for thirty years and then release them selectively, and the Official Secrets Act makes it a crime for anyone to reveal, without authorization, any information acquired by him as a public employee, contemporary histories of Britain are typically undocumented, vague, lame, and opinionated or, when they are memoirs, self-serving. Such legal prohibitions on the flow of information are obviously significant in a country where doctors and nurses are public employees. These restraints necessarily render all data suspect. Where poverty and unemployment are endemic, educational attainment is low, and pollution is uncontrolled, as in Britain, there is no reason to be much impressed by statistics indicating a high average life expectancy, for example.

At the same time, the secretiveness of the government and the potency of the laws restraining the press assure that newspaper versions of events will be more than fair to the government. Most of the information I use has passed through a filter of official approval, simply by virtue of the workings of the Official Secrets Act and the government's exercise of prior restraint, and because of regular, off-the-record briefings of journalists by government, which are a major source of news. The information to be found in the British press is alarming enough, however incomplete it may be, to provide material for a dozen sobering volumes. It is absorbed by the public very quietly, which means that the government has made a fair estimate of public passivity. In fact, the greatest burst of official admissions about the scale of contamination and the fecklessness of Sellafield's management arose just at the time that the planned expansion of the plant and the construction of a similar plant at Dounreay, Scotland, were made known. Why this should be an effective way of managing public reaction I cannot speculate, but the government has encountered no important resistance, so, within the

British context, the government must be seen to have handled it all very smoothly. Admissions of incompetence seem to ingratiate, and to enlist loyalty in their public. Admissions of past mismanagement seem to be accepted as earnests of good intention. American opinion was, of course, unruffled. So, again, all this alarming news alarmed no one, the plant grows and prospers, and the little earth is still without an ally, despite all the funds and clubs and naturalists, or because of them.

This book is essentially an effort to break down some of the structures of thinking that make reality invisible to us. These are monumental structures, large and central to our civilization. So my attack will seem ill-tempered and eccentric, a veering toward anarchy, the unsettling emergence of lady novelist as petroleuse. I have had time and occasion to note the disproportion between my objective and my resources. If I accomplish no more than to jar a pillar or crack a fresco, or totter a god or two, I hope no one will therefore take my assault as symbolic rather than as failed. If I had my way I would not leave one stone upon another.

I am angry to the depths of my soul that the earth has been so injured while we were all bemused by supposed monuments of value and intellect, vaults of bogus cultural riches. I feel the worth of my own life diminished by the tedious years I have spent acquiring competence in the arcana of mediocre invention, for all the world like one of those people who knows all there is to know about some defunct comic-book hero or television series. The grief borne home to others while I and my kind have been thus occupied lies on my conscience like a crime.

This book is written in a state of mind and spirit I could not have imagined before Sellafield presented itself to me, so grossly anomalous that I had to jettison almost every assumption I had before I could begin to make sense of it. My writing has perhaps taken too much of the stain of my anger and disappointment. I must ask the reader to pardon and assist me, by always keeping Sellafield in mind—Sellafield, which pours waste plutonium into the world's natural environment, and bomb-grade plutonium into the world's political environment. For money.

PART ONE

The first questions that arise in attempting to understand Sellafield, and more generally the nuclear and environmental policies of the British government, are: How have they gotten away with so much? and Why on earth would they *want* to get away with it? To put it in other terms, why should the relationship of those who govern Britain to its land and population be that of a shrewd adversary contriving to do harm for profit? For decades the British government has presided over the release of deadly toxins into its own environment, for money, using secrecy, scientism, and public trust or passivity to preclude resistance or criticism and to quiet fears. Such extraordinary behavior cannot have a motive in any usual sense, since it is in no one's interest. It has, however, an etiology and a history, in which the institutions which expedite it and the relations it expresses evolve together. This is of more than casual interest to Americans, because there is no stronger cultural force than atavism. Our past is a good commentary on the future we seem to be preparing for ourselves.

It is often said that Britain has no written constitution.

If a constitution is a body of law that defines the fundamental relations among the elements of a society, then Britain has an ancient one indeed, solidly encoded, enshrined in literature, in history, and in an array of institutions. The core of British culture is Poor Law, which emerged in the fourteenth century and was reformed once, in 1834, when it became the Victorians' notorious New Poor Law. It remained in force until 1948. Then it was superseded by the Welfare State, in which its features were plainly discernible.

In essence, Poor Law restricted people who lived by their labor to the parish where they were born, and mandated assistance from the parish for those who were needy and deemed deserving of help, while wages were depressed to a level that made recourse to such help frequent. This often meant entering a poorhouse, institutions whose wretchedness made them, over centuries, objects of the minutest study to generations of philanthropists. Working people who were forced to accept parish assistance, and whose destitution was absolute, and who were found otherwise worthy of aid, surrendered whatever rights they may have had. Or the fact that they had no rights was thoroughly and ingeniously exploited once they accepted this status. Under the Old Poor Law, before the 1834 reforms that made the operation of the system more punitive and severe, child paupers, that is, the children of destitute parents, were given to employers, each with a little bonus to reward the employer for relieving the public of this burden. The children would be worked brutally, because with each new pauper child the employer received another little bonus. To starve such children was entirely in the interest of those who set them to work. Aside from all the work the child performed under duress, its death brought the reward that came with a new child. The authorities asserted an absolute right to disrupt families, and to expose young children to imprisonment and forced labor. The invasiveness of the Poor Laws was never impeded by the development of any system of assured legal rights, with which the entire institution would have been wholly incompatible and out of sympathy. Leslie Scarman, a member of the House of Lords and a legal authority, has written: "It is the helplessness of law in [the] face of the legislative sovereignty of Parliament which makes it difficult for the legal system to accommodate the concept of fundamental and inviolable human rights."* More to the point, the social history of Britain has never reflected any sense of the unconditional value of human lives or any respect for the modest baggage of person and property, the little circumference of inviolability on which personal rights depend.

The indigent who were considered worthy of parish assistance were called paupers. The unworthy, those who were considered able-bodied but shiftless, were not to be relieved, though in fact they were often assisted on the same terms as the "deserving poor," that is, meagerly and punitively, since the system was in any case preoccupied with the need to withhold charity, considered the great source of moral corruption of the poor and therefore the great source of poverty. So late and well reputed a social thinker as the young William Beveridge urged that starvation be left as a final incentive to industry among the shiftless poor. Beveridge was to become the father of the Welfare State.

The mandate of Poor Law charity was only to provide

---

* Quoted by Ralf Dahrendorf, *On Britain* (1982), p. 124.

subsistence, because if the recipient of charity were to do as well as the independent worker, the worker, too, would become demoralized and slide into pauperism. At the same time, a very important article of economic faith was that the wages of workers could not exceed subsistence—if they did, the depletion of capital would cause a decline in investment and employment that would return the worker unceremoniously to something less than the level of subsistence. So it was difficult to make the situation of paupers less desirable than the situation of the employed, especially considering the horrendous conditions under which most work was done. Paupers were subjected to the miseries of the separation of their families, and they were auctioned off or forced into emigration, depending on the improvisations of local authorities determined to keep relief recipients to an absolute minimum. To assure that parish assistance would be limited to those who were qualified by birth or legal settlement to receive it, the movement of workers was narrowly restricted.

The abusive treatment of paupers was justified on the grounds that it discouraged the class above them, the employed, from sinking into Poor Law dependency, and it was justified by the suspicion that the class below them, the poor unworthy of assistance, were to be found among them despite all precautions, and it was justified on the grounds that dependency easily became habit, that charity demoralized its recipients. Every worker was a potential pauper, and every pauper was a burden, presumptively demoralized, and an agent of demoralization of others. These assumptions created and sustained the legal situation of the great majority of British people.

Even now British subjects have no rights established

in law. Supposedly they enjoy customary rights, but where in their harrowing history any custom friendly to their interests could have been established I am at a loss to know. Until the 1980s people were imprisoned without trial on the word of the arresting officer for appearing to intend a crime, under the Vagrancy Act of 1824. Such arrests have supposedly ended, but in Britain few things ever really end. Clearly, those against whom such laws are carried out have none of the protections we imagine to be "Western." The laws are highly consistent, however, with the conditions of the Poor Law, which voided every notion of individual rights—except, of course, the slippery right to subsist, always boasted of and worried about despite high rates of death among paupers, and at best enjoyed only by those the parish could or would relieve.

Parliament, which would be the political expression of fundamental rights in the British population if it were a straightforward representative institution, is characterized by an extraordinary mix of prerogatives and disabilities, which combine to weaken all other institutions without creating real power in the Parliament itself. No British court can override the laws it passes. Recently it abolished seven major elected city governments, including the Greater London Council, an action of special significance because these were major power bases for the Labour Party. To this day Parliament can expunge an official crime by legalizing an action of government retroactively. It has almost perfect legislative freedom, in theory, but in fact it has no right to any information the Prime Minister does not give it. A bill, while it is in preparation, is an Official Secret, forbidden to Members of Parliament as to anyone else. Experts on modern Britain describe the system as an elective dictatorship, but I have my doubts about that formula. Whitehall, the bureaucracy that is actually charged with developing legislation, collecting information, and implementing policy, is not elected, and does not change with parties or ministries. Question time, when the Prime Minister replies to questions from Members of Parliament, can deal only with specified subjects. It is forbidden, for example, to inquire into purchases made by the National Health Service. As Ralf Dahrendorf, head of the London School of Economics, says in his book *On Britain*, "What happens is not decided in Westminster [that is, in Parliament]. It is not even discussed at Westminster in any detail. As a result, the visible political game becomes strangely superficial." Mr. Dahrendorf admires the system, on balance. However, that the "deceptively lively adversary surface"* of parliamentary activity *is* a surface, rather than an authentic and consequential process of deliberation, means that even the right to vote is a very small concession of power on the part of those who do decide "what happens." In other words, there is a pervasive absence of positive, substantive personal and political rights in Britain.

The structures of institutions express conceptions of society. Sellafield amounts, in its dinosaur futurism, to a brutal laying of hands on the lives of people: a blunt, unreflecting assertion of power. It is the same unchallenged assertion of economic prerogative that legally immobilized the majority of the British population for five hundred years, so that the cost of relieving their wretchedness, when wretchedness became extreme, could be contained.

The movement of workers from the country to the

---

* P. 102.

cities and from the North to London demonstrates that
these laws, which mandated the forced return of strayed
workers to the place of their legal settlement, were not
consistently enforced, though in the beginning of this
century Beatrice Webb, guiding spirit of Fabian socialism, may be heard grumbling about the numbers who
were returned and the costs entailed. If cities represented opportunity, relatively speaking, migrants had
the impetus of destitution and humiliation at their
backs, to enhance the pull of urban life. Everywhere,
whatever they did, workers were seen as burdens, actual
or potential, and this perception governed every aspect
of their existence.

America had its paupers and poorhouses, through
the nineteenth century at least, though these institutions
seem never to have seized on the national imagination,
or to have remained in the popular memory. Haw-
thorne's failed utopia, Blithedale, becomes a poor farm.
Thoreau is visited at Walden by paupers, including one
who declares himself "deficient in intellect." In the early
twentieth century, Booker T. Washington notes that
blacks are almost never paupers, by which he means that
they are provident and self-sufficient people. Given our
profound cultural debt to Great Britain, it is no wonder
if our policies with regard to the poor are sometimes
crude and high-handed. But imagine what it would be
like if we had truly replicated British social organization,
if every American who lived by a wage had been
immobilized to simplify the administration of welfare in
the event he should need it, and if this arrangement had
been persisted in for hundreds of years. This would
surely trench very far on the dignity and liberty of
citizens, and their pursuit of happiness.

Margaret Thatcher is easing the burdens of Britain by

cutting back on education, health care, and other
threadbare amenities, pinching one of the poorest populations in Europe, supposedly to punish and cure their
poverty. This is a typically visceral reaction against the
supposed cost to the state of allowing meager comfort to
people perceived as demoralized and reduced to dependency by intemperate generosity.

Such patterns of reaction are as old as Poor Law itself.
William Beveridge, who wrote the celebrated Plan for
the Welfare State in 1943, promised "subsistence" to
employed people in conditions of high national employment. This is the great socialist dream against which the
present government recoils. Americans, perhaps because they romanticize their origins, never think of the
lives of British people as circumscribed and poor, historically or at present, though many of them are descended from the outcasts and refugees of this same
penurious system. That any society could promise so
little, and then renege, seems preposterous, except
against the background of British social history.

How does one back away from a promise of subsistence? As an economist, Beveridge knew that the historical meaning of the word has only been that one
should not die of starvation in its starkest form. Disease,
poisoning, exposure, malnutrition, and exhaustion have
never been treated as incompatible with subsistence,
though they have slain multitudes. In other words,
though in Britain historically it has been used to establish the wage of a worker, in theory and in practice, and
as the measure of mercy to the afflicted, and as the
animating vision of the armies of reformers, subsistence
has always been, conceptually speaking, a rotten nut.
Beveridge's promising it under certain—historically
atypical—conditions implies that, under other condi-

tions, people must expect less. If subsistence seemed to the British public of the late forties a bright prospect, less-than-subsistence must have seemed to them to describe their situation before the war. (During World War II, food rationing improved the British diet and health. Since Beveridge presided over the distribution of food, perhaps a definition of subsistence was inferred from his successes in relieving poor nutrition. However, postwar austerities returned nutrition to prewar standards.)

It should shock us that the virtual constitution of a modern state could imply that the subsistence of its people was not to be assumed in the event of unfavorable, and highly typical, economic conditions. William Beveridge wrote an escape clause into his Plan for the Welfare State, and Margaret Thatcher has availed herself of it. Beveridge, claiming the influence of Maynard Keynes, opined that government could stimulate employment so as to maintain it at the level necessary to make his plan viable. In other words, in the absence of ongoing government action, the plan would collapse. For years the British government has systematically created *un*-employment, so the plan, whatever it amounted to, has lost its economic rationale. It was designed to be possible only if it was not especially necessary.

The ancient pattern of dubious charity provoking horrified reaction against the object of charity is being repeated now in the radical attack on the meager fabric of public amenity—they are selling the Thames—and, in general, on the standard of living of the poor, whose consumption of medical care has been curtailed by the collapse of the National Health Service, and whose real income has been reduced by the curtailment of every kind of provision the state so gallantly undertook in

1948, appropriating to itself thereby, as William Beveridge knew and said, the socialization—the control, that is—of demand.

The mechanism built into the British Welfare State which allows demand to be depressed was perfected in the Poor Law system. It is the combination of poverty-level wages, heavily taxed for good measure, with a system of national decency and, shall I say, comity which brings real income back up to the level considered economically convenient by the government of the day. Britain has never had a minimum wage. Wages have always been notional. Against the large reductions of public provision now being made, a few percentage points of increase in wages claimed by the present government mean absolutely nothing.

What is happening now is a counterattack against the demons which the British ruling class and middle class have always felt to be released by charity, whether public, private, secular, or religious, and also by prosperity among those classes of people for whom prosperity is not customary. To these supposed erosions of order and value, and infringements on the wealth of the well-to-do, the characteristic response is a swingeing punishment.

Poor Law and the society it generated amount to a prehistory of America, not only because its mechanisms of expulsion peopled these shores, but also because it created the legal context which made life, liberty, and happiness revolutionary aspirations. The early history of Poor Law is barbarous, and its violence will seem specific to an early period. But the occasion for this book is the knowing and calculated contamination, by the British government for profit, of a populous landscape, with the most toxic substance known to exist on earth. And it is my impression that leukemia, like older misfortunes, alters one's appearance for the worse. If anyone wishes to object that my comparison is unfair or sensational, I reply that the motive behind all these martyrdoms is profit, and, more precisely, the threat and terror of redundancy, of lives existing in excess of economic demand. Without Sellafield there would be even more unemployment. Barbarous exactions are still being made on economic pretexts.

Poor Law appeared first in the form of the Ordinance of Labourers promulgated in 1349 under Edward III, which required work at legally limited wages of all able-bodied workmen and workwomen "free or servile," anyone who refused being jailed "until he find security to serve in the form aforesaid." The occasion for this ordinance was the Black Death, which had depleted the population so severely that workers were in great demand, and accordingly able to ask for higher wages than they had previously received. There was at the same time an inflation in prices which must have made higher wages necessary, because working people chose to be idle rather than to accept the pay they were offered, even though they were, as described in this same law, people with no resource but the sale of their labor. Clearly the interest of the state, and its authority, merge with those who employ. The principle established in 1349, and not departed from even now, when unemployment is created as a policy of government, is that "the commonwealth," the employing minority, has a presumptive right to the labor of working people, with no obligation to acknowledge its value, whether as

established by demand or as giving consideration to the
share of labor in the creation of wealth.

In the theory of political economy, workers compete to
sell their labor in a free market. In theory, which bears
a most complex relation to practice, their labor creates
all value. The Ordinance of Labourers is directed
against the development of a labor market in conditions
which would make demand favorable to the workers'
interests. The more typical condition of "redundancy,"
of glut in the labor market, would be allowed to cheapen
labor, however, though low wages, by driving women
and children into employment, contributed greatly to
the excess. Read aright, Poor Law is a system which
severs work from any notion of its objective worth by
criminalizing idleness. This unconditional claim made
in the Ordinance of Labourers on free and servile
alike—on those who had managed to wrest themselves
from serfdom and those who had not—raises questions
about the distinctions between free and enslaved workers which remain lively into the twentieth century.

The Ordinance of Labourers contains another feature
characteristic of later Poor Laws. It forbids charity to
"sturdy beggars" on the grounds that such people are
guilty of withholding their labor. The adjective "sturdy"
implies that the old or infirm may be relieved, enforcing
the distinction on the charitable thus:

And because many sturdy beggars, so long as they
can live by begging for alms, refuse to labour,
living in idleness and sin and sometimes by thefts
and other crimes, no man, under the aforesaid

penalty of imprisonment, shall presume under colour of pity or alms to give anything to such as shall be able profitably to labour, or to cherish them in their sloth, that so they may be compelled to labour for the necessaries of life.

Idleness in the able-bodied is wicked and leads to wickedness, and therefore to extend charity to the undeserving is itself an act worthy of punishment. Anyone confronted by a beggar who might be called able-bodied would exercise caution. Thus, at the small cost of denying alms to some who deserved them, the great public benefit would be gained of extracting labor from those fit to work. Sound morals and sound economics at a single stroke.

It is characteristic of Poor Law as a phenomenon to attempt to suppress the charitable impulse and, where benefit is transferred, to maximize its effectiveness as social coercion. The conditions imposed on the giving of charity make it no charity at all. That for which it is exchanged—submission, in a word—makes it instead an unusually good bargain, especially since more niggardly assistance is more effectively controlling. This fortunate conjunction of advantages will be sought with increasing rigor and system, but by the same methods, over centuries. Beatrice Webb herself will brood over the well-being of the working poor with a sublime concern that they should not be corrupted by any largesse, public or private, that succors them when their need is not exquisite. In fact, a most rigorous discrimination between worthy and unworthy poor, morally earnest in the extreme but sadly inclined to pinch and humiliate the worthy in order that no reprobate should escape unpinched and unhumiliated, will become the primary

care of British philanthropy, enlisting the efforts of the
finest spirits and the loftiest minds.

Certain features of this fourteenth-century ordinance
should be noted. It deals with working people exclusively, depresses their wages and exacts their labor, and
worries over their tendency to be taken for or treated as
needy people, unable to work, whose right to charity is
implicitly conceded. For a long time, well into the
twentieth century, the words "poor" and "labourers"
and "workers" will be used interchangeably. Like the
Ordinance of Labourers, the Poor Laws will be directed
at working people, whose normal condition is assumed
to be poverty. Paupers, the destitute, those who fall
short of subsistence, are simply workers in sickness or
old age or widowhood or madness or despair, or whose
trade has become obsolete or whose industry has gone
into crisis, or whose wages have fallen so low that they
work and are still indigent and dependent. "Pauper" is
simply Latin for "poor," logically enough. The word
implies a distinction whose reality is doubtful at best,
since the whole class of workers or poor were governed
by laws supposedly designed to relieve and discourage
pauperism.

The distinction between workers and the poor is not
made, because workers *are* poor, and, as a class, are
vulnerable to utter destitution. This fundamental relationship of labor to the purchasers of labor will attract
rationalizations the way a magnet does pins. Frequently
these rationalizations are at odds with one another, but
this does not matter, because as justifications of an

existing order, their very variety indicates profound consensus.

The Ordinance of Labourers, with its refusal to distinguish between free and servile, was clearly designed to shore up erosions in the feudal system. An act of the twelfth year of Richard II, in 1388, reinforced this effect. This law anticipated the later Acts of Settlement, which would continue in effect in association with the Poor Laws down to 1948. It forbade any laborer to leave the town or borough where he lived without a "letter patent containing the cause of his going and the time of his return." Towns were to maintain stocks where any laborer traveling without a letter could be kept "until he have found surety to return to his service or to serve or labour in the town from which he comes." Those who cannot work, "beggars unable to serve," are immobilized in the same way and by the same means. Thirty years after the plague, labor was still scarce, still exacted: "As well artificers and craftsmen as servants and apprentices" are "to be forced to serve in harvest at cutting, gathering, and bringing in the corn." The law specifically limits the wages of categories of workers, "because servants and labourers will not and for [a] long time have not been willing to serve and labour without outrageous and excessive hire and much greater than has been given to such servants and labourers in any time past."

This law is often described as important because it makes a distinction between the idle and the truly needy. However, it does nothing of the kind. It restricts the movements of the impotent poor in order to restrict more effectively the movements of the able-bodied, who begged and wandered just as they did. It is designed to make unnecessary the distinction required by the ban on charity for sturdy beggars in the Ordinance of Labourers. Again, while the Poor Laws are always treated as if they were intended to alleviate poverty, these early statutes from which they derive are clearly designed to immobilize labor and bring down its cost. The features of the laws affecting the impotent are designed to ensure these results.

The interpretation of the laws as primarily charitable provision for the needy, rather than as attempts to control the cost and supply of labor, has given great currency to the view that they originated in the breaking up of the monasteries by Henry VIII. The eighteenth-century writer Frederick Eden, in his classic work, *The State of the Poor* (1797), traces the Poor Laws to these origins, as do Adam Smith and Karl Marx, Carlyle, Disraeli, Hilaire Belloc, and others. Marx mentions the fourteenth-century laws in his discussion of the Poor Laws in *Capital*, but he makes nothing of them. Occurring as they do long before the Reformation, they are anomalous, if one assumes that the intent of the laws was to create a secular equivalent of the disrupted institutions of Christian charity. The miseries of the mass of rural people driven from the land described in More's *Utopia*, published in 1515, make very clear that the problems of poverty and beggary were fully present when the monasteries still flourished. That More should not have mentioned them in association with the suffering he describes hardly encourages one to believe that they figured significantly in alleviating need.

The institution threatening to collapse in the fourteenth century was not monasticism but feudalism. The

Black Death had ravaged the society, transforming the situation of the workers by making them few. The great restlessness of the people during this period issued finally in the Peasants' War. Jean Froissart, a contemporary of these events, a Frenchman with a courtly bias, reports in his *Chronicles* that in 1381 there occurred "great disasters and uprisings of the common people, on account of which the country was almost ruined beyond recovery." Says Froissart: "It was because of the abundance and prosperity in which the common people then lived that this rebellion broke out." Other historians find a cause in the Ordinance of Labourers. The high wages which workers seem to have been able to command where the law was skirted, together with the oppressiveness of its enforcement where it was not, might very well have produced a revolt like Wat Tyler's. "These bad people," wrote Froissart, "began to rebel because, they said, they were held too much in subjection, and when the world began there had been no serfs." Serfdom was at that time, according to him, especially widespread in England.

The peasant rising swept through the country and entered London. The great army of the poor was finally dispersed by the precocious statecraft of the young King Richard II, who granted their request, the end of serfdom. When they had returned to their villages, he sent men at arms after them, the letters freeing them were torn up and scattered in front of them, and their leaders were killed, more than fifteen hundred men. Seven years later, in 1388, Richard II legally forbade the movement of working people over the countryside. The advantages of the law as a system of social control are obvious. Further, it prevented workers from finding the best market for their labor, keeping wages low even at the cost of creating idleness and indigency. This immobilization continued as a feature of the Poor Laws for five centuries. Clearly the legislation is not based on any charitable design. The intention of this law is to assure that the poor remain poor.

Other acts leading up to 43 Elizabeth, as the classic version of the Poor Laws passed in 1601 is known, are unabashedly ferocious. They lay claim to the labor of working people with sovereign indifference to the questions of freedom such appropriation necessarily entails. In the first year of the reign of the child king Edward VI, it was enacted "that if any man, or woman, able to work should refuse to labour, and live idly for three days, that he, or she, would be branded with a red hot iron on the breast with a letter 'V', and should be adjudged the slave, for two years, of any person who should inform against such idler." A glimpse of hell, surely. "And the master was directed to feed his slave with bread and water or small drink and such refuse meat as he should think proper; and to cause his slave to work, by beating, chaining, or otherwise in such work and labour (how vile soever it be) as he should put him into." Furthermore, "if he runs away from his master for the space of fourteen days, he shall become his slave for life, after being branded on the forehead or cheek with the letter 'S'; and if he runs away a second time and shall be convicted thereof by two sufficient witnesses, he shall be taken as a felon, and suffer pains of death, as other felons ought to do."

The preamble of the statute makes explicit the philosophical grounds for the criminalization of idleness: If "the vagabonds who were unprofitable members, or rather enemies of the commonwealth, were punished by

death, whipping, imprisonment, and with other corpo-
ral pains, it were not without their deserts." A vagabond,
for the purposes of this law, is a man or woman who has
been idle for three days—not in itself proof of wanton-
ness or evil disposition. That a transgression so inconsiderable should bring down such ruin upon a man or
woman accused of no crime except the withholding of
labor means that the claim of "the commonwealth"
upon the work of working people was, in effect, without
limit. If "the commonwealth" failed to assert its claims in
the person of an informer—the framers of this law were
aware there existed in this world "foolish pitie and
mercie,"

> yet never the less justices of the peace shall be
> bound to inquire after such idle persons; and if it
> shall appear that any such have been vagrant for
> the space of three days, he shall be branded on the
> breast with a 'V', made with a hot iron, and shall be
> conveyed to the place of his birth, there to be
> nourished, and kept in chains, or otherwise, either
> at the common works on ammending highways, or
> in the service of individuals after all such former
> condition, space of years, orders, punishments for
> running away, as are expressed of any common or
> private person to whom such loiterer is adjudged a
> slave. If vagabonds [that is, those reported to have
> been idle three days] are carried to places, of which
> they have falsely declared themselves to be natives,
> then for such lie they shall be marked in the face
> with an 'S', and be slave to the inhabitants or
> corporation of the town, city or village where he
> said he was born in, for ever.

This law was soon repealed, though there were subsequent attempts to revive it. Its provisions, while extreme, nevertheless anticipate features of later laws.

A series of Elizabeth I's laws made wages variable to deal with the problem of the changing prices of "things belonging to servants and labourers," because earlier restrictions "could not be carried into execution without the great grief and burden of the poor labourer and the hired man." A statute of 1572 required "a general assessment, or tax, for the relief of the impotent poor," any remaining money to be used to employ rogues and vagabonds under supervision. The term "vagabond" was this time carefully defined to include, among others, "bearwards and common players in interludes." If Shakespeare had not been allowed to dress as a great man's servant, he would have fallen under the provisions of these laws, at a great ultimate cost to the tourist industry.

Another law passed in the reign of Elizabeth established "houses of correction," where "youth might be accustomed and brought up to labour, and then not like to grow to be idle rogues; and that such as be already grown up in idleness, and so rogues at this present, may not have any excuse in saying that they cannot get service or work." By the same authorities, needy persons were to be supplied with wool, hemp, flax, and so on, so that they could, of course, work.

A statute of 1597 established four overseers in each parish for setting poor children to work. The law, addressed to the problem of vagrancy, directed parishes to establish housing for their own impotent poor.

These laws create an increasingly coherent system in which wages are fixed to the cost of subsistence, the movement of the poor is stabilized, the young poor and

the idle poor are compelled to work—at textile manufacture, interestingly. And in being so employed, they
are all objects of charity, or, to couch the matter in
the modern and secular terms more appropriate to it,
they are burdens on the taxpayer. Eden says, "The
situation of the Poor even after the passing of the 43
Elizabeth, is represented by some authors as exceedingly
deplorable; and the assessments for their relief are said
to have been so low that many perished for want," and
that "The salaries of the masters and governors [of
houses of correction] were directed to be paid by the
treasurer of the Poor; and these alone must have added
heavily to the county charges." Never mind. The moral
high ground won here will never be relinquished. Ever
afterward the poor will be probationers while they work
and reprobates when they cannot work, though there
will be no obligation on anyone's part to employ
them.

Indeed, during the period from the Black Death
onward, more and more land was being given over to
the pasture of sheep—first of all because there were too
few people to keep the land under cultivation, and then
because wool became an important trade while England
supplied the textile industry of Flanders, and then, in
the sixteenth century, because England established its
own textile industry.

It might seem that the ultimate pole of expropriation
has been reached when one has only one's labor to sell.
A further extreme is reached when the principle is
established that one's labor has no objective value, not
even of the kind established in a market. One last
dispossession remains. The process was called "depopulation," and it involved the pulling down of towns and
villages—not the passive decline of the rural economy,

but the active expulsions of people who had ceased to be of economic use.

The practical disadvantages that attended the consequent poverty and disruption, though never sufficiently great to outweigh their advantages, were noted. Attempts were made to slow or reverse this change, which was nevertheless precipitous. An act passed under Henry VIII in 1533–34 describes the amassing of pasture in a few hands, the destruction of towns, the driving up of the cost of food by the decline in farming, "by reason whereof a marvellous multitude and number of people of this realm be not able to provide meat, drink and clothes necessary for themselves, their wives and children, but be so discouraged with misery and poverty that they fall daily to theft, robbery and other inconvenience, or pitifully die for hunger and cold." An act passed under Elizabeth in 1597–98 attempts to protect husbandry and tillage, on the grounds of their being "the occasion of the increase and multiplying of people both for service in the wars and in times of peace, being also a principle means that people are set on work, and thereby withdrawn from idleness, drunkenness, unlawful games and all other lewd practices and conditions of life," and these being the means by which "the greater part of the subjects are preserved from extreme poverty in a competent estate of maintenance and means to live." Yet during the period in which the poor were being driven off the land, laws were made which vehemently criminalized idleness, wandering, and begging. The contradiction did not go unremarked. A speaker in the House of Commons in 1601 is reported to have said, "If we debar tillage, we give scope to the depopulator; and then if the poor being thrust out of their houses go to dwell with others, straight we catch them with the Statute of Inmates; and if they wander abroad they are within danger of the Statute of the Poor to be whipped." The displaced wretches Thomas More described as thronging the roads of sixteenth-century England were redundant, because the economy had undergone its first major revolution, and their skills as agricultural laborers existed in excess of economic need as the landscape became in effect industrialized, reorganized to supply wool for manufacture. Become unprofitable, they had the ground taken from under their feet, and they were compelled to wander, in violation of law, and to beg, in violation of law, and to live in idleness, in violation of law. How brutally people as frail as these surely were would have been affected by the beatings and brandings and like horrors to which their circumstances exposed them, a moment's reflection will suggest. Imagine a father or mother of young children—these laws were gender-blind—starved and filthy and bewildered, beaten bloody for the crime of resourcelessness and then driven into the road again.

The Tudors are known to have executed many thousands of these "felons," a word which, as we have seen, can merely signify the readiness of King and Parliament to punish vagrants as if they were criminals. But the outright executions cannot represent any significant fraction of the process of liquidation these depopulations must have entailed. Since life was precarious and child mortality rampant under the best circumstances, and since these people were transient and fugitive, their perishing may not have presented itself to their contemporaries as a phenomenon of demographic importance.

But only imagine—this was at the beginning of the period of enclosure of the commons, that great, bold movement of privatization which swept away the ancient right to keep a goat or a goose on common ground. It was superseded by the commercial interests of the landowning classes, the agricultural laborer being denied the only resource he had besides his redundant strength. The loss of the commons meant that the rural worker was conceded no independent, customary place in the world. His subsistence was nothing in the balance against the profits of the proto-industrialists, the suppliers of wool to the textile makers, Flemish or English.

The Poor Laws have always overlapped and contradicted one another. People who can neither stay where they are nor go elsewhere are in trouble. The question is simply whether, when, and how it would be advantageous to punish them. Fundamental institutions of British society were thus formed around legislation unconstrained by any conception of individual rights, not solicitous enough of those they affected to make their obedience possible, or even to allow for their survival. What the laws did do indisputably was to give a free hand to whoever wished to enforce them, whoever felt his interest to be infringed by this nuisance population.

I will make, at this point, a very rude suggestion. Why were the commons enclosed? To make room for sheep. But were the common lands sufficiently large, in a sparsely settled country, to be needed for that purpose? Even though penury and sharp dealing are pervasively characteristic of the British ruling class, still, would the addition of some hundreds more sheep be a sufficient motive in itself for an action so catastrophic, to these men of great fortune, whose portfolios were becoming even then increasingly diversified? A possible second motive would be "depopulation" itself. Their wages and the common lands together provided the whole subsistence of agricultural laborers. While neither was sufficient by itself, the loss of both would surely hasten the disappearance of the economically redundant. The word "redundant," which Americans take to be a euphemism for "unemployed," actually has a long and savage history, denoting an excess population, one whose sufferings prove it should not exist, a notion with many notable applications; for example, in Social Darwinism and in eugenics. To conceive of others' lives in such terms is chilling, expressing a hostility to their hopes and interests deeper and more intractable than ordinary hatred.

The English and Scots countryside have long had emptiness as their primary ornament. They are seen as unspoiled by time, though in fact it was industrialization that created all that emptiness, as surely as it created Liverpool and Manchester. It seems Wordsworth used his influence as Poet Laureate to keep the railroad out of the Lake District, so that the region would remain inaccessible to working people from the cities. Whatever Cumbria has suffered, care has been taken to spare it one affliction.

A law promulgated under King Edward VI provided that if a poor woman gave birth to a child in a parish where she had no settlement, she was to be beaten and imprisoned for six months. The Poor Law specified that an indigent person should be provided for by the parish where he was born. A vagrant woman with child therefore exposed any parish through which she passed to possible liability for the support of an infant during the whole of its life. To avoid the potential addition of even one mouth to the burdens of a parish, this extraordinary transgression against decency was considered worthy of the seal of a prayer-book-writing little king. To see the mighty thus bend out of heaven, as it were, and touch these most precarious lives makes me think of Gloucester's line from *King Lear* about flies and wanton boys. One may suppose that cosseted men, brooding over the injury done to them by indigency among the laboring classes, might have come up with a law like this one, not quite thinking through to the effect it would have on the survival of woman and child. But this supposition seems charitable.

I think it is at least as probable that the death of child or mother was not an unacceptable outcome. Henry Fielding quotes Edward VI with admiration to the effect that vagabonds are "spittle and filth" to be expelled from the body of the state because they have no use. There was nothing so cheap as the lives of redundant people. This was true from the beginning of the Tudor period to the end of the nineteenth century. How the statement should be amended now is a question that will arise in due course. For the moment I wish only to point out the re-enactment of this early revolution from above, when the economy was changed to produce mass unemployment, which was then condemned and penalized for the fact of its own existence, and made the pretext for undermining the relief system which supposedly protected against the consequences of unemployment. Exactly this has happened in postwar Britain, redundancy this time created not by the industrialization of agriculture but by the abandonment of industry.

*   *   *

The aversive reflex against the supposedly charitable aspects of the Poor Laws has been an extraordinarily important force in the development of British culture and society. Landed proprietors were obliged to pay poor rates for laborers who lived on their lands—which seems fair enough, since the workers were a burden on the taxpayer in the exact proportion that the proprietor chose to stint on wages. But where these lordly personages were concerned, fair enough was never good enough. By the simple device of pulling down the cottages on his land, or letting them rot from neglect, the proprietor made his workers find shelter elsewhere, and excused himself from the obligation to pay their rates. As one unhappy consequence, laborers were obliged to walk miles every day to the fields. As another, they became burdens on ratepayers other than the employer who then could profit more substantially by stinting their wages and by turning them away when they were not needed. He could save the expense of sustaining them in sickness or slack times and then have the benefit of a reasonably intact work force when it was wanted. Instructed by the example of landlords, neighboring villages limited the building of cottages on the theory that they were thereby limiting the numbers of potentially indigent who could settle in them. But since these populations were obliged to live as near as they could to their work, the result was simply a fantastic crowding of existing cottages, for which exaggerated demand made rents high and repair unnecessary.

Having neither time to cook food nor fuel to cook it with, farm laborers bought what they ate from shopkeepers, like industrial workers. Their cottages leaked, but they had no way to dry their clothes, which they wore till they rotted away. Human waste is often described as being in heaps beside their cottages, and this compounded the effects, in terms of ill health, of the crowding together of malnourished and exhausted people. Robert Hughes, in his book *The Fatal Shore*, about the penal settlements in Australia, observes that convicts seem not to have found life on the farms there worse than in rural England. The wretchedness of life in England established the norms of life in Australia, where English wretches went to be punished.

The importance of the ideas that idleness should be regarded as a crime and that charity corrupts by encouraging idleness cannot be overstated. Conceding everything one must about the hypocrisy and corruption of church-administered charity, the kind prevalent in Europe into this century, still the transaction is sanctified, words of consecration have been said over it, and there is nothing in writ or tradition to suggest that any soul, however disreputable, who comes to the table of charity eats and drinks to his own damnation. In England, however, just such reprobation is believed to follow any undeserved relief. The moral deterioration set on by charity predisposes the worker to the vices that produce indigency—in other words, suffering is the fault of the poor, liable to be exacerbated rather than relieved by any effort to help them. Misery itself becomes a proof that its sufferers are indulged and lacking in character—and there has always been enough misery in Britain to demonstrate, by this reasoning, prodigious generosity toward a public that is always less deserving. Beatrice Webb, my favorite British socialist, never wearies of warning against the "demoralization" and "pauperization" which may follow from any brush with public relief. There is, therefore (so great is the tendency of charity to corrupt), a presumed obligation to withhold relief even from the worthy. Much is always made, in British thought, of the need to distinguish between the "deserving" and the "undeserving" poor, but the institutional history of the Poor Law system will make it clear that the only way to deserve help is not to need it, or at any rate not to ask for it. Those who ask to be assisted are not merely therefore suspect but also exposed to the risk of decline into the condition of unworthiness they might to that point have escaped. At the same time, those who might choose to starve with their families rather than accept relief on such terms are viewed as deserving of imprisonment on those grounds. In the whole abundant British literature the Poor Laws have generated, hardly a kind thing has been said about them—except, of course, that they are the folly of a too melting nature and that they anticipate the Welfare State. The most persistent criticism made of them is that they create poverty, the same sad result that Frederick Eden laid to religious charity before the destruction of the monasteries.

They did create poverty, of course. In every form, their effect has been to depress wages—by imposing them legally, or by preventing workers from seeking to sell their labor at its market value; or by criminalizing idleness, not merely in men, but in women and children also, obliging them to labor simply to remain unmolested; or by subsidizing wages to bring them up to the level of subsistence, relieving employers of even the practical need to maintain their workers at the level of "physical efficiency," while exacting labor as proof of meriting such largesse.

Evolution has given the accolade of stability to the sharp tooth, the thick skin, the small brain. Poor Law theory plods on through volatile centuries, only more itself, losing reflection to instinct. If one was inclined to believe that ideas over time acquire greater delicacy or complexity, the history of these laws would constitute a refutation. Herbert Spencer, the nineteenth-century theorist of Social Darwinism, is no advance on Frederick Eden, or William Beveridge on William Hazlitt: reflections on the Poor Laws, among that select group whose thoughts are recorded, are always critical—saddened, indignant, or resigned. Every criticism of the system that can be made has been made at one time or another. But its assumptions are never called into question—or they were once, by Adam Smith. Smith made the novel case that the wealth of nations should be calculated in terms that included the prosperity of their working people: "No society can surely be flourishing and happy of which the far greater part of the members are poor and miserable. It is but equity, besides, that they who feed, cloath and lodge the whole body of the people, should have a share of the produce of their own labour as to be themselves tolerably well fed, cloathed, and lodged." He went off to his grave with praise ringing in his ears, and was not seriously attended to, then or since.

The assumption that workers must be poor passes unmodified into the literature of political economy. The "labor theory of value," the idea that labor produces all value, makes its appearance very early, in the work of the seventeenth-century writer William Petty, and quietly establishes itself as orthodoxy. It seems never, however, to imply—except to Smith—that the laborer, the producer of wealth, would have any share in it. On the contrary. The poor, being the producers of this valuable commodity, labor, rather as sheep are of wool, must be kept in an optimum state of productivity. That is, they must be obliged to work in order to live. If they get a little money ahead—this wisdom is often repeated—it goes to drunkenness and rioting. And in any event, the political economists discovered "wage-fund theory" and "subsistence theory," which meant together that only a certain portion of the national wealth can be spent on wages, beyond which the whole would be reduced, and that wages tended naturally to sink or rise to the level called "subsistence." Science (for so they took these theories to be) frequently obviated certain questions of justice, while throwing others into sharper relief. By the light of these theories, for example, it was plainly to be seen that the prosperity of one worker could come only at the cost of other workers, so equity required what fate decreed, that wages should remain low.

Again, invariably the interests of the state, and its authority, merge with those who employ. Much is made of the polarization of classes in Britain. Its origins are not mysterious. Until 1948 the working class was governed by a restrictive legal code which did not touch its socioeconomic betters, those prosperous enough to have their idleness called leisure. At the time of the earliest statutes there was as yet no compulsory provision for the poor which would make their indigency an expense to the taxpayer. The loss of labor would affect only employers. Still, the laws make the idle worker "an enemy of the commonwealth." Again, the enforcement of the law depends upon informers, whose reward could be the vagrant person himself or herself in the role of slave. To keep a slave would have been at least as costly as to hire someone when wages meant only subsistence, so the law clearly assumes that any informer would be of the employing classes. The law is written to assure unobstructed access to the work of laborers by potential employers of labor. The law seems designed to settle the question of whether the laborer owns his labor as property and has the right to govern its use. He or she has no such right. The punishment for such an assertion of freedom is slavery.

Lacking the right to withhold their labor, or to sell it in the best market, the poor were utterly vulnerable to what Karl Marx calls "exploitation." One would be hard put to find a better word. The early laws teased loose any connection between work and payment. Subsequent laws put charity in the place of pay, insofar, at least, as wages were subsidized by a system designed to compensate for their meagerness, intermittency, and downward drift. One worked to stay out of the clutches of this charity if at all possible, and to be found deserving of it if all else failed. It was this William Blake must have had in mind when he wrote: "Charity would be no more/If we didn't make somebody poor."

It seems strange, in retrospect, that the persistent problem of poverty should vex the best minds of England for so very long. In 1704, Daniel Defoe launched a distinguished tradition in a searing attack on the Poor Laws, addressed to the Parliament, called *Giving Alms, no Charity*. He argued against the employment of the poor in workhouses and houses of correction on the grounds that poverty derives from the "crimes of our People," which he enumerates: (1) luxury, (2) sloth, (3) pride. The English are well paid but improvident. "There's nothing more frequent, than for an Englishman to Work till he got his Pocket full of Money, and then go and be idle, *or perhaps drunk,* till 'tis all gone, and perhaps himself in debt." This is why "children are left naked and starving, to the care of the Parishes." Defoe argues that the labor of the dependent poor will cause economic dislocations and "take the Bread out of the Mouths of diligent and industrious Families to feed Vagrants, Thieves and Beggars." Defoe has put his finger on a problem, of course. The competition of forced labor would lower the value of wage labor. His solution is based on the assumption that people are indigent through their own fault, that the rigors of the law should force vagrants and beggars to find the work which is, he insists, available.

We have all seen people grow warm denouncing the chiselers and the spongers, either strong or fat, who squander their food stamps on soft drinks and corn chips, while their unkempt, innumerable children wait on the curb. They have earned their corn chips many times over in savings to the public treasury, since their mere existence, whether real or rhetorical, has always counseled restraint.

Defoe's tract, however, is interesting because it is an early example of the erecting of economic theory on a highly peculiar conception of labor. His language makes no distinction between the independent poor and the indigent or dependent poor, since a "general Taint of Slothfulness" predisposes the entire class to improvidence and beggary. Yet, he declares, "even all the greatest articles of Trade follow, and as it were pay

Homage to this seemingly Minute and Inconsiderable Thing, *the poor Man's Labour."*

The great prosperity of England, its "vast Trade, Rich Manufactures, mighty Wealth," rests, ironically, on this most uncertain foundation. If a man who gets a little money ahead uses it only to buy drink, to lie in the alehouse while his children starve, high wages are in no one's interest. Workers in other countries earn less, he says, and live more comfortably. The situation, as Defoe sees it, is this: Poverty is caused not by too little money but, in the short term at least, by too much of it. The disposition of the poor toward sloth and luxury means that any excess of money might plunge them into ruin. Money has just the same destructive effect as idleness, into which it is readily converted. Defoe claims there are a thousand fathers of families "within my particular knowledge" who "will not work, who may have Work enough, but are too idle to seek after it, and hardly vouchsafe to earn anything more than bare Subsistence, and Spending Money for themselves." A class of such extreme moral fragility, at the same time so crucial to the national well-being, needs not charity but regulation.

Defoe's essay is an early application of Poor Law thinking to the new circumstances of industrialization. It is an attack on the "charitable" aspect of the laws, which were devised to exact labor but which critics from Defoe to the present would accuse of impeding access to labor by corrupting the working class. The discourse is a dialectic of frying pan and fire, centered around an unquestioned assumption that the poor are in need of aggressive management for their own well-being, which altogether coincides with Britain's commercial success. Industrialism took the form that it did because rural populations were driven off the land into a world that harrowed them for their misery. The factory system throve on the existence of a class without resource or expectation, a stigmatized class whose existence at worst and at best was penal servitude. If this class had not existed, industrialization might have occurred differently, not only in Britain, but in every country where Britain served as an example. Defoe's essay, written at the very start of the eighteenth century, already describes England as a trading and manufacturing country, and already expresses fears of foreign competition. (The Muscovites are acquiring British technology, and there people work for "little or nothing.")

Appearing this early, in a setting where feudalism had changed rather than receded—to dispose of people so peremptorily is a great demonstration of power, not a renunciation of it—and where feudalism was put on guard repeatedly, and never overthrown, it is to be expected that certain features of the old order should be retained in the new industrial society. Defoe was aware that the wealth of the country was expanding rapidly, and that these changed fortunes were the result of the development of the textile industry begun by Queen Elizabeth. How is the new wealth to be distributed? Will there be a proportional rise in the prosperity of all ranks of society? Defoe's tract is an argument for keeping the vast class of labor on a short tether, a subsistence wage. The workhouse was, after all, the least controversial element of the Poor Law system, savoring little of charity in the Scriptural sense, while it enforced the all-important role of worker upon its inmates and, if it was managed properly, turned a little profit. Nevertheless, the system does, as Defoe represents it, increase the proportion of the national wealth consumed by the poor by excusing them from the need to be provident. His wastrels spend themselves into poverty and then become dependents of the parish. Where their wages, if they were frugal, would have sufficed, they have consumed their earnings as well as whatever they and their children end up costing the taxpayer. Aside from its other inconveniences, he argued, the Poor Law system makes the poor secure.

Henry Fielding also wrote about the Poor Laws, and submitted a plan to Parliament for their reform. He was warmly in favor of workhouses, and wished only to make them more efficient. Fielding was astonished, as a great many writers would be, that "in a country where the poor are, beyond all comparison, more liberally provided for than in any other part of the habitable globe, there should be found more beggars, more distressed and miserable objects, than are to be seen throughout all the states of Europe." Among these "miserable objects," however, those unable to work were so few that they should be left to private charity, and the Poor Law system designed to give work to the able-bodied. Streamlined according to his recommendations, the poorhouses would be considerably more profitable—off-loading the lame and the blind would necessarily effect a savings.

Not surprisingly, Fielding has a theory of wages, which is linked to his grand design thus: Wages should be fixed, to discourage idleness. This reform would defeat those who, "if they cannot exact an exorbitant price for their labour, will remain idle." It will provide magistrates with proof of the willingness to work, or its opposite, for purposes of distinguishing the idle from the incorrigibly idle. Again, work is pried loose from pay. Work proves one deserving—more effectively when the issue of willingness to work is not obscured by the possibility of holding out for a higher wage. In

Fielding's scheme the workhouse is already integrated into the wage system, since to qualify for non-punitive accommodations there, one must have been employed.

The economic and moral argument that wages must and will be kept low is embodied in the work of the earliest English socialist, the cotton manufacturer Robert Owen. Owen built a model factory community called New Lanark, which, through new housing, communal cooking and laundry, schooling of children, and programs of recreation, elevated the living standards of his employees. In an introduction to his *New View of Society*, he explicitly describes factory workers as human machines among inanimate machines "which it was my duty and interest so to combine, as that every hand, as well as every spring, lever and wheel, should effectually co-operate to produce the greatest pecuniary gain to the proprietors." Visitors and dignitaries the world over came to admire his success.

In a book entitled *Observations on the Sources and Effects of Unequal Wealth* (1826) the American socialist L. Byllesby noted dryly that Owen paid only the standard factory wage and realized a healthy profit. He objected that it was still required of "the producers to surrender a part of the avails of their labor to those who hold a claim of proprietorship over the necessary means for putting their labours to use." Aside from its interest in establishing a sense of the terms of political discourse in America and before Marx, this criticism is a fascinating early example of the American tendency to miss the point entirely where British social reform is concerned. Owen was making a demonstration of the fact that, properly supervised, rationalized, and instructed, working-class lives could be lived decently and wholesomely at market wages. His experiment tended to prove the old wisdom that the "labor fund"—the sum of money considered to be available in the economy to be paid out in wages without inhibiting the creation of capital, thereby diminishing total wealth—was sufficient to provide adequately for workers. Truly what Owen attempted was the minimum of change, in effect a vindication rather than a reform of the capitalist system—and by capitalist I mean exactly what Marx meant, a system in which a working class is exploited to produce wealth in which they have no share, a system which considers subsistence an appropriate compensation for the mass of people, and an appropriate condition of life.

A partner in New Lanark was Jeremy Bentham, who devoted much time and thought to designing a perfect pauper asylum. It was, of course, a workhouse—which still meant "factory." Bentham's scheme would find labor suitable to the varying powers of its inmates, so that the very old or young or ill could make themselves useful. He proposed that children born to inmates be detained into their early twenties—that is, through their peak earning years. Altogether, he felt confident that he could turn pauperism into profit. He dreamed of a chain of these asylums, built on a plan that would facilitate supervision of work and policing of morals, as well as the combining of duplicated work, such as the cooking of meals. He reasoned that such good order would create positive happiness, especially among those who, lacking experience of the outside world, could not make comparisons. This vision of philanthropy does not seem to me remote from the system Bentham realized in partnership with Owen. Both dream of creating a circumstance in which profit and happiness will be maximized together, a sort of transfiguration in which the factory system will be revealed in glory. Certainly

the condition of Bentham's paupers approached nearly
enough to the conditions of his and Owen's employees
to demonstrate how very fine a distinction it was that
separated the great class of the poor into two great
categories, worker and pauper, whether as objects of
philanthropy or of punishment. In fact, to distinguish
between them is usually an error. *A History of Socialism* by
Thomas Kirkup (1906) reports that five hundred of the
two thousand workers at New Lanark when Owen came
there in 1800 were children "brought, most of them, at
the age of five or six from the poorhouses and charities
of Edinburgh and Glasgow." In other words, the exploitation of the work of child paupers of which Bentham
dreamed was already carried out on a very significant
scale. It is startling how often British philanthropists
have dreams which if they were to wake they would find
true.

The American Byllesby criticized New Lanark on the
grounds that it was a profit-making scheme and that it
was "directed primarily, to the better formation of
human character, and secondarily to ameliorating the
condition of the labouring or productive classes."
Byllesby feels this is putting the cart before the horse,
that "if pecuniary concerns are first put in a train of
amendment, or reform, the human character will, of
itself, keep an equal pace in the expansion of its amiable
traits, and suppression of its evil capabilities." But,
again, Byllesby has missed the point. His assumptions
are the individualist kind, for which we have long been
notorious. Owen has no place in his system for "the
human character"; it is the *worker* he is concerned with,
a being wholly defined by his class status and his
economic function. Owen's object is to fit him humanely
to a role he must occupy in any case.

Owen's experiment is said to have inspired no imitators in Britain. This remark is misleading, since it implies that the project was a true innovation and that subsequent practice bore no relation to it. Owen set about to stabilize workers in their condition as workers, that is, to proof them against the vices and accidents that created illness, misery, and degeneracy, by regulating the particulars of their existence. It was simply a repetition on a smaller scale and in a more sanguine spirit of the great British project of balancing the working class on the knife edge of subsistence. The minimum of national wealth allowable to the working class was the amount that maintained them in health and vigor. Failing this standard, they fell into illness, despair, and indigency, and became unproductive and, compounding every evil, a charge on the taxpayer. The preempting (exactly the right word in this context) of such small choices of food, drink, and use of leisure as were available to workers encroached far on the small freedom of unenfranchised people. But Owen did not object to actual slavery. Byllesby's assumption that human nature would flower spontaneously if economic injustice were ended is not a variant of Owen's idea but its opposite.

From 1838 to 1848 the Chartist Movement, a huge and truly popular call for specific political reforms—universal manhood suffrage, annual parliamentary elections, secret ballot, equal electoral districts, removal of property qualification for members of the House of Commons, and pay for its members—made its long transit across the skies and vanished. It was important because it expressed the views of the mass of people about where their salvation lay. Thomas Carlyle, often listed with Robert Owen among the great social critics of the era, heard Chartism as an inarticulate groan, its true meaning being, "Guide me, govern me! I am mad and miserable and cannot guide myself!" In his view it signified the need of the great mass of people to be truly governed by their natural superiors. He argued that the "free Working-man" should "be raised to a level, we may say, with the Working Slave . . . Food, shelter, due guidance, in return for his labour." In typically febrile language, Carlyle expressed the view that carried the day—minus, of course, slavery, which shocks the modern reader, but which had plausible humanitarian arguments to be made for it over and against the situation of the British working class, who were utterly destitute when they were of no economic use to anyone, and who were without resource from the day they became too ill or frail to market themselves as labor. That a change in the legal status of powerless people should be expected to so revolutionize established norms of conduct toward them simply expresses Carlyle's positive feelings about slavery, which he elsewhere likens to matrimony.

But just to the extent it might be assumed that slavery involved some obligation to sustain the worker in sickness as well as in health, in infancy and in old age, slavery would be a bad bargain. To the extent that the purchaser of labor has an economic interest in the well-being of the laborer, he is restricted and encumbered in the matter of working conditions, for example. And to the extent that he can purchase labor as a commodity, abstracted from whatever vulnerable creature yields it, of the best available quality and in the quantity that suits his needs from day to day, he has a vastly cheaper labor source than slavery. Therefore, Carlyle must make his case for slavery on humanitarian grounds.

Modern American commentators take Carlyle to be decrying the crimes of capitalism, which means *our* crimes, and the inhumanities of *our* interrelations. Sensing rebuke, they fall to tugging their forelocks and, thus occupied, are too busy and too happy to wonder about the sort of social criticism that would idealize slavery (though, in fairness, it should not be assumed that these commentators have in fact read Carlyle, rather than merely having acquired a phrase from him in the course of their education).

Carlyle, like the less objectionable Owen, is an example of the most characteristic feature of British thought; that is, the tendency to criticize in such a way as to reinforce the system which is supposedly being criticized. The essay *Chartism* is, first of all, a defense of the New Poor Law, whose draconian provisions Carlyle laments roundly—before putting his blessing on them. He is telling his wealthy audience that they should utterly disregard the political demands of the lower orders, that these demands have a significance just the opposite of their intended meaning, that they are a cry not for greater freedom but for less, and that they in fact justify a vastly greater intrusion into the lives of poor people than the New Poor Law itself accomplished. Quite hilariously, he denounces laissez-faire as a doctrine which permits the working class to do as it will, and freedom, he says, has been its ruination. In the face of the undeniable misery which has been the consequence of the stewardship of the governing classes, and in the face of popular demand for an expansion and reconstruction of representative government, Carlyle argues—to the sort of audience sure to be receptive—that England is unhappy because the naturally superior have not governed with sufficient thoroughness.

This is more than flattery—although it is most certainly flattery, too. It is a call to reform, of a kind taken very seriously, not least because it represented so small a departure from established practice. Regulating the poor had been the great preoccupation of British social thought from its precocious beginnings, always with the same object, securing ample labor at a favorable price. In defense of the severity of the New Poor Law, Carlyle argues that everyone should work or die—and that the poor should take it as an honor that their lives will be the first to feel the force of this great principle. Slavery is one logical extension of the sort of thoroughgoing, personal governance, uncomplicated by redundancy, of which, in Carlyle's view, the mass of men stood desperately in need.

British social thought may as well be imagined as occurring this way. It takes place in a country house built and furnished to accord with conventions polished by use, a house filled with guests, great and minor luminaries, ornaments of literature, the sciences, the church, and of philosophy and politics. Most of them, not coincidentally, are cousins at some remove. They are charmed to find in one another just that streak of intuitive brilliance they had always admired in themselves, and to be confirmed in their sense that they are true members of a group in which there are no impostors, by a very great similarity of taste, of interest, of sympathy. It is a leisurely visit, some centuries in length, and in due course everyone has confessed his weakness for Hesiod, and admired the garden, and regretted the weather. The evenings would perhaps have begun to weigh, if someone had not suggested a game called Philanthropy. The rules of this game are very simple. One must justify things as they are by attacking things as they are. It is a philosophic game, perfectly suited to showing off a fine wit. It has even the thrill of risk, since it invites subversive ideas. But the point is always, of course, to achieve a resolution that will bring the argument right back where it began.

This distinguished party warms to the challenge. And how affecting it is to hear them, one after another, in the language of statesman and moralist, decry the sufferings of the poor, until it seems that the very table they sit around must be made into splints and crutches and the topiary garden planted in potatoes. Then, just when the pleasure of participation in this virtuous fantasy is at its height, that is to say, just when the temptations of virtue are most intense, then the player reveals the illusion: This "virtue" is not virtue at all, but an evil to be scrupulously avoided. A little thrill of relief passes over the company when their world is safely restored to them. But the risk is never as great as it may seem. Any strategy is sufficient in defending the moon from the wolves.

It is a distinguished company, and everyone seems willing to hold up his part in the game. Daniel Defoe, Bernard de Mandeville, Henry Fielding, Adam Smith (who did not understand the point of it, and was given a hearty cheer and sent off to bed). There is no need to observe chronology, since at this table Jeremy Bentham might find himself seated by Beatrice Webb, and Herbert Spencer by John Stuart Mill. This is only to say that their reflections on this subject accumulate rather than develop, in the manner characteristic of rationalizations. Their disputations produce a welter of harmonious contradiction, the sort of thing that happens when any argument is welcome that will prop a valued conclusion. So the centuries pass.

The influence of this genteel assembly can hardly be overstated. Only consider how important the notion of excess population—basically the artifact of an odd and unsavory history—has been to Britain, and therefore to the world. Malthus felt he observed the fact of population being restored to equilibrium with food supply in the misery of the poor, but at the time he wrote, the importation of wheat—bread was the food of the British poor—was restricted, and land had been converted to pasture which had formerly been used for growing food, and both industrial and agricultural workers had lost access to independent means of subsistence, the first by being crowded into urban slums where there was no corner of open land, the second by being crowded into rural slums where no bit of land was conceded to them. Social reformers early in this century wrote dreamily of the little garden plots of Belgian workers, who throve better on, of course, lower wages than their British counterparts. But the British laborer had no little plot of land. Irish immigrants shared quarters with their notorious pigs, which they slaughtered for food, but that was considered degraded. In fact, British workers, rural and urban, died of exclusive dependency on a meager wage, made up in part, especially among farm laborers, of parish relief, more parsimonious because it was paid by ratepayers rather than employers, and because, being "charity," it always remained discretionary. The relation of population growth to the productivity of land, which Malthus tidily but meaninglessly described as increasing geometrically in the first case and arithmetically in the second, had nothing to do with the misery and vice he set out to account for. His was merely an early instance of the tendency to refer the consequences of a remarkably artificial situation to the hard laws of nature.

We have never ceased to talk about overpopulation, though true instances of it seem very rare. The English workingman Francis Place, having contrived to educate himself under astonishingly unfavorable circumstances, became the first writer in English to argue for birth control. He accepted Malthus's view that workers were poor because there were too many of them, and he argued that their improvement lay in self-restraint. Of course, like poor people almost anywhere, they had children for their economic value. As late as the present century the prosperity of a family fluctuated with the number of employed people in it, and the early redundancy of the father, as well as such vicissitudes as sickness and injury, were more easily borne the more shoulders they fell on. Any glut of labor was the result of employing people from childhood, for sixteen hours a day or more, at wages that denied them any possibility of withholding their labor. A higher wage would have relieved the glut by allowing women to stay home with their new infants, or allowing families to keep older children at home to attend to younger children. Physical efficiency would have been enhanced at the same time.

But "working class" is the primary term in British social thought. The coercive implications of the phrase are glossed in every version and institution of the Poor Laws, right through Fabianism and the Welfare State, which is only the latest version of the ancient view that what the worker earns, his wage defined as subsistence, is not his by right, as property. The welfare system indeed *assures* that the wage will never amount to any specific sum of money but will be nuanced to provide subsistence itself, not in a money equivalent, however calculated. In other words, workers earn their existence by working. At the same time, their employment exists at the pleasure of those who employ, first of all, the government. The Welfare State, being designed for the employed, is designed for the less necessitous. Those who do not work are historically regarded with a sort of scandalized aversion, exactly like people without caste in a society based on caste.

Francis Place accepted blame for poverty on behalf of working people, and accepted the notion that there could be an excess of people relative to their economic usefulness, and that that excess should be eliminated by sexual abstinence so that it need not be carried off by disease and starvation. He accepted the notion that there was a natural balance, a marketplace of survival glutted with laborers.

Darwin was likewise indebted to Malthus, and freely acknowledged his influence. Overbreeding relative to available food sources harrowed out those less well adapted to survive, in Darwin's view. Herbert Spencer, who was to Beatrice Webb as Aristotle to Alexander, seized upon the idea immediately as a model drawn from nature, which justified just such horrors as had been Malthus's point of departure one hundred years previously. Competition supposedly described the hard-scrabble existence of the laboring classes, who sold their labor by the day and were constantly thrown out of work by a change of season or fashion, or the invention of a new machine. It was a commonplace that they helped one another as best they could through these disasters. The idea nevertheless had great impact because it made death a legitimate part of the social and economic order, a function rather than a malfunction of political economy, a measure of the extent to which the new industrial society cleaved to the ways of nature rather than departing from them. It affirmed the idea that there existed a human surplus, whose survival could only be secured at the cost of creatures worthier to survive. In such a context subsistence would be a positive reward, easily withdrawn, as in Darwinian nature.

The penchant for developing theories to account for suffering and death proved useful. Press and parliamentary reports of the Victorian period describe a system of exploitation which ravaged the culture, extruded every ounce of labor from those able to work—including small children and women recently delivered—starved them, beat them, poisoned them, housed them in cellars into which sewage drained from the streets, taking from them everything that could be taken, turning every good thing into a mode of exaction.

Housing, for example, which was typically wretched, was often rented to workers as a condition of employment, so that the employer could withhold rent and therefore be at no risk of losing it, raise rent as a control on the amount received by the worker in wages, and expel the worker from the house when he was fired, adding to the terrors of idleness the loss of shelter, and dreaded vagrancy. The houses would be inhabited, however foul or wet or crowded, and no one would be in a position to request repairs. How could more economic advantage be found in the supplying of shelter? Well, this sort of housing permitted the claim to be made that the factory was a relatively wholesome environment for young children, and the sixteen-hour day a positive philanthropy. It was not merely dirtiness or neglect but low morals as well that were said to characterize the working-class home, whole families and even lodgers sharing a single bed, where there was a bed. This state of affairs seized upon the imagination of moralists, who lamented all the degradation such a life must entail. At the same time, the morally convenient view that degraded people were less sensitive to insult and imposition and therefore could be exposed to conditions their betters would find intolerable resulted in savings to both employer and taxpayer.

The Victorians, who presided over this squalor at its worst, evolved a fastidious system which seems, in context, designed to widen the chasm between the wealthy and the poor by devising a system of virtue dependent on privacy and on elevating amenity to the status of morality. A Victorian lady was confined from the time her pregnancy became discernible. A working-class woman worked until she delivered, gave birth in a crowded room, and returned to work a few days afterward. At the end of her ten to sixteen hours, her clothing would be sodden with milk. When morality was so thoroughly confounded with mere fastidiousness, these women must have seemed appalling indeed, and their children as well, who could scarcely pretend to any innocence at all, as their betters understood the word. Added to this compromising proximity of family members was a virtually absolute lack of sanitation, which involved the laboring classes in an awareness of the particulars of bodily processes as far beyond the pale even as begetting and birthing. Their quarters were swamped in excrement. Again this was not a peculiarity of urban life, since the depopulations of the countryside had squeezed farm laborers into rural slums entirely as dense and filthy as industrial slums. All this simply indicates how invasive and pervasive were the exploitations of the controlling classes, who determined every particular of the lives of those who lived by labor.

Reading Taine or the earlier St. Simon or Zola, one can see an equivalent in another context. There is no reason to imagine that any European country experienced less severe exploitation of the masses of its people than Britain did. But British society was remarkable for the progressive nature of its immiseration—extreme poverty in association with unprecedented advances in science and technology, and in national power and wealth. It was Britain's inspiration to transform static feudal custom into dynamic capitalist system, to convert wealth into capital; that is, to release its astonishing powers of self-replication. Italian city-states and the Low Countries developed banking and trade to a very high point. But Britain added industrialism, with its special opportunities for the growth of capital through expansion and innovation. Britain's importance as a center of trade and finance gave it world markets, highly suitable for absorbing its production of staples such as textiles, which were cheap because they came from modern factories whose operatives were as badly paid as any in the world. I am aware of the little charts which show a worker of whatever kind earning however many shillings and pence. Much relevant social history is devoted to turning out the pockets of these people, and graphing their contents for comparison. This is meaningless, however, when workers were required to rent their shelter and buy their food from employers, or from others who profited from the necessity of their living near the factory where they were employed sixteen or eighteen hours and more, or from the need to buy food daily and close at hand—which made workers the primary victims of adulteration as well as high prices. To this must be added the inflation of the costs of necessities which can always be enjoyed by monopoly suppliers. And then employers had the uncontrolled right to fine for defective work or damaged equipment. All this makes a wage itself meaningless, since by one means or another its real value is susceptible to reduction at any time. Health conditions and length of life are better indicators of how one is repaid for one's labors. In industrial cities the average length of working-class lives was seventeen years. For reasons touching the profitability of enterprise no wage could buy the mass of people a breath of fresh air or a taste of clean water. They subsidized with their health and lives the profits of industrialists.

It is a seamless history. The contamination of modern Britain with radioactivity is done by industry, for profit. The health of those affected is an appropriation of property, since, as Adam Smith argued (and Marx seconded him, as he did often), one's labor and therefore one's ability to labor is a property and a patrimony. To deplete strength and life is to overstep limits no one should be required to cross. This is an ethical argument, of course, though couched in economic terms. People have always sold their lives along with their labor. To refuse has rarely been a practical option. That is why the American abolitionists, well before the Civil War or the *Communist Manifesto,* treated chattel slavery and wage slavery, in those terms, as one phenomenon. The fulcrum of the competitive system of capitalism in Britain was the worker and his wage, though this fact reflected peculiarities of the social order rather than any real economic necessity. Britain had, after all, so many varieties of competitive advantage that higher real wages, which would be repaid in a larger domestic market, could have been absorbed, while the loss of discipline enforced by desperation would be compensated for by greater vigor and longer lives among the industrial work force—a body whose skills, virtually unique in the world at that time, were among Britain's great economic advantages and therefore deserving of consideration, if on economic grounds alone. The uneconomic side effects of the degradation of the work force are as obvious now as they were then, and yet it has persisted, under the pretext of economic necessity.

Marx was correct in seeing society as roiled to the bottom by "economic relations," which amounted to the confiscation in whole or in part of the peace, health, beauty, dignity, strength, and span of life of the great majority of people. Just such a high-handed and unembarrassed claim to dispose of others' lives as was made by industrialists, by agriculturalists, by slave traders and colonizers (both in routing indigenous populations and in dumping unwanted Britons and Irish in remote corners of the earth), is expressed in the conduct of government and industry in Britain today. Now, as ever, there is no sound calculus of interest that will justify, for example, the importation of nuclear wastes in vast quantities by a small country which has presumed to deal with them because its specialists have shrugged off the difficulties involved, choosing financial gain against the risk, if not in fact the certainty, of unthinkable loss. It is a strange abuse, from which financial benefits can be expected only if it is assumed that there will be no unconcealable catastrophe, that other countries will not be shamed out of paying for this disreputable service, that the country will never face boycott or ostracism on account of Sellafield, that its impact on the national health will never become insupportably grave.

I suggested earlier a moral polarization in British society, which was economically based in the sense that it gave moral value to the refinements of life made possible by privacy and amenity. This is not to imply that working-class people were uninjured by the lack of these things. It is only to point to the pattern of a morality grown increasingly "nice" while the lives of most people conditioned them to circumstances this morality would find degrading and presumptively immoral. The conditions which yielded all the imagined and actual vice and the scandalous absence of proper hygiene were created by the very class who could hardly bring themselves to speak of them. The accelerated progress of refined sensibility in the opposite direction from physicality and squalor suggests that goodness is being defined as the opposite, not of evil, but of poverty.

The Victorians are famous for their rigorous notions of female purity, their exquisite awareness of female delicacy, their lisping attentions to pretty children. Yet industrial labor from its beginning was overwhelmingly the work of women, and of children as well. Women were not recruited into a system designed for endurance greater than theirs. The worker around whom industry *developed* was a woman. Men, when Engels wrote in the nineteenth century and when Beveridge wrote in the twentieth, were considered unemployable except at odd jobs by the time they were forty, and regularly became the dependents of their wives and children, who, though preferred to men as workers, were paid considerably less (which heightened, one must assume, the degree to which they were preferred). The primary earner in a household having so brief an active life, the incomes of families were pieced together out of the trifles paid to supposed dependents. All in all, the economy of the nineteenth century was as if designed to demonstrate the toughness of women, while at the same time the myth of female delicacy elaborated itself endlessly.

Again, I am suggesting here a morality of denial, which reached an especially high degree of development and has been put to especially rigorous use, in Britain. I do not think that historically the British upper classes really have not known how the working class lives, though (then, as now, and at frequent intervals between) revelations of poverty were greeted with little shrieks and exclamations such as "Among us! In this land so widely admired for its decency and civilization!" The proof of awareness, highly specific however deeply suppressed, is precisely this mirror-image reality, in which everything is reversed. Sheltered children, sheltered wife, dominant father/husband, leisure, freedom from financial dependency, access to the countryside, the ability to observe a highly nuanced propriety requiring a variety of dress, an array of forks and glasses—all these things were absolutely not characteristic of working-class life. The Victorian home flourished over and against the working-class household in which everyone above the age of five or six might be employed, leaving the littlest ones untended through an endless Victorian workday. "Slum" is cant slang from the word "slumber." These people must have done little more than sleep in the few hours they had to themselves. It was often remarked that they were deficient in domestic culture.

*   *   *

Defining one's values in opposition to the conditions of life of poor people has numerous advantages, especially when merely to admit to a knowledge of such conditions is compromising. One can experience the difference between oneself and those less fortunate as the difference between virtue and vice, and that is comforting. It reinforces the old faith that poverty is the consequence of a degraded character. A more efficient justification of an existing order can hardly be found than the notion that those who enjoy its advantages are, in fact, better, not through any special attainment, though these are extravagantly admired where they occur—but simply as the repository of a particular experience.

The "clever" of Britain, whose distinguishing marks are verbal first of all, consider themselves their culture's ornament and justification. Therefore they are very poor critics of the system that has created them or, more precisely, that has decreated the skeptics and competitors who might have dashed their self-confidence or the confidence of others in them. It is the convenient faith of Britain that it is a pure organic growth, whose gifts are for referring great questions to custom and intuition. This belief peripheralizes thinking. Among the whole class of the verbally clever there is a fluency which is social rather than intellectual—though the tendency of the culture is to suppress this distinction. Beatrice Webb declared herself "the cleverest member of one of the cleverest families in the cleverest class of the cleverest nation in the world," and her faith made her prolific. It also led her to retail and recodify the pet theories of that cleverest class and nation. These theories, whether

capitalist or Malthusian or Social Darwinist, have always
had the public good as their first object. The problem,
of course, is one of definitions. For example, historically
and at present the British have seen little benefit in wasting education on the non-clever. Malthus, weary of being seen as an enemy of the poor, advocated education
as a means of encouraging religion and sexual abstinence among them. William Hazlitt dismissed his suggestion, pointing out that workers in the North were
often literate and religious and it only made them un-
ruly. Cleverness has always been a rationed commodity.

Since cleverness is overwhelmingly class-associated,
certain habits of thought, angles of vision, and styles of
articulation have authority without reference to their
implications or their consequences. Naturally any reform of the institutions of society will reflect the thinking
of its best minds, its "cleverest class," and this is certainly
the case with the British Welfare State. This accounts for
the tenaciousness of primitive social ideas, such as the
positive value of class itself, and the tendency of the poor
to be corrupted by the alleviation of their poverty, and
the assumption that the state, in accepting any practical
responsibility for the general welfare, has taken on a
killing burden, and must decline, and should be honored
and venerated in its consequent mediocrity—that is, in
the shabbiness of public provision which, on a sort of
lifeboat analogy, is taken to reflect the precipitous decline
in value of aid extended too generally. There is such a
profound bias against generosity in British culture that it
is entirely possible for them to argue that where misery
is achieved a too melting generosity must have lain behind it. It has been characteristic of British social theo-
rists, from Defoe to Malthus to the writers of the New
Poor Law to Herbert Spencer to the Webbs to Mrs.

Thatcher, to cry out for an end to generosity on humanitarian grounds.

All this moaning and groaning creates the impression
that some dreadful sacrifice is being made, the pelican is
plucking flesh from its breast to feed its children, and
these children, cosseted wretches, never see any point in
leaving the nest. As a matter of objective fact, ordinary
British people have always enjoyed a very small portion
of the wealth of the British community. Subsistence has
always been considered an appropriate description of
well-being as it applied to the general British public. It
is the noise these people make—the mandrake groans,
the Carlylean tirades and tears—that has led the world
to believe a powerful spirit of justice was at work on that
island, like Jehovah at Sinai, producing out of darkness
and thunder the very tablets of righteous law.

This great coherency of theory and practice should not
be taken to imply that the British world picture could
not accommodate another vision of society and economy. As I have said, Adam Smith argued that the wealth
of nations should be measured by the productivity of
their people, which was enhanced by liberal wages and
by relaxation, even "dissipation and diversion," and by
education, to moderate the stultifying effects of the
division of labor. Capital accumulated through monop-
olies and other policies that depressed consumption
while raising profits therefore diminished wealth. Foreign trade and manufacturing had created new opportunities for the powerful to engross profits, rather than
to share them, and Smith remarks, "All for ourselves,
and nothing for other people, seems, in every age of the

world, to have been the vile maxim of the masters of mankind."

Why, or by what means, Adam Smith has been made to seem the apologist for capitalism *par excellence* I cannot tell. The economy of colonialism, mercantilism, and monopoly which he criticizes at such length clearly corresponds to Marx's capitalism, in which the worst potentialities Smith described are fully realized. In *The Wealth of Nations,* Smith used the wealth of the American colonies, elsewhere called their "happiness," to demonstrate his argument that high wages were consistent with low prices and with productivity. He compares the prodigious growth of population in the American colonies, for him the sign as well as the source of wealth and productivity, with Europe and Britain, where "it is not uncommon, I have been frequently told, in the Highlands of Scotland for a mother who has borne twenty children not to have two alive."

Marx repeats, more explicitly, the distinction Smith had made almost a century before between an economy designed to promote the concentration of capital in the hands of merchants and industrialists and one characterized by the welfare of the general population, also using America as the example of the second type. While the distinction these writers make is not without complication, it is not particularly subtle either. And yet it has been wholly lost.

Chapter 33 of *Capital* is an attack on E. G. Wakefield's book *England and America,* published in 1833. Marx is emphatically determined to establish a distinction between capitalism, an economic system in which the working class is wretched and dependent, and its "direct anti-thesis." Wakefield describes America as retarded in its development by the high cost and status of labor. He calls the wealth of Americans "capital," and Marx replies:

> Political economy confuses on principle two very different kinds of private property, of which one rests on the producers' own labour, the other on the employment of the labour of others. It forgets that the latter not only is the direct anti-thesis of the former, but absolutely grows on its tomb only.

He declares that the United States is not a capitalist country. He describes the economic conditions that prevailed here as in fact an "anti-capitalist cancer." America has this character as a colonial economy, but except for a glancing allusion to Australia, it is his single example, the foil against which he defines capitalism by contrast. He opposes the American social order to capitalism on the grounds that "the means of production and subsistence, while they remain the property of the immediate producer, are not capital. They become capital, only under circumstances in which they serve at the same time as means of exploitation and subjection of the labourer." The last words of *Capital*—the "Come quickly, Lord Jesus"—of this supposed fountainhead of socialist thought, are "Capitalist mode of production and accumulation, and therefore capitalist private property, have for their fundamental condition the annihilation of self-earned private property; in other words, the expropriation of the labourer."

In Marx's view, the laborer in America escapes expropriation because he can acquire land, basically. But the essential thing is not the availability of land but the value of his labor, the fact that he can sell it for a wage that significantly exceeds subsistence. "So long, therefore, as the labourer can accumulate for himself—and this he can do so long as he remains possessor of his means of production—capitalist accumulation and the capitalistic mode of production are impossible." Risking hubris, I must say that Marx seems to have things backward here, since accumulation in the first instance is the result of a high wage, and wages of course go to those who are not possessors of means of production. (If he means to include money among the means of production, as it would be entirely reasonable to do, then of course I withdraw my objection.) As he says a little later, in America "the wage-worker of to-day is to-morrow an independent peasant or artisan, working for himself." In other words, the value of his labor allows him to become a possessor of means of production—*not*, as Marx makes very clear, a capitalist, however, for the same reason that his own employer was not, in Marx's terms, a capitalist. The demand for labor precludes "the social dependence of the labourer on the capitalist," that indispensable requisite of capitalist production. Marx says, "In the Colonies property in money, means of subsistence, machines and other means of production, does not as yet stamp a man as a capitalist if there be wanting the correlative—the wage-worker, the other man who is compelled to sell himself of his own free-will. Capital is not a thing, but a social relation between persons, established by the instrumentality of things." The "man . . . compelled to sell himself of his own free-will" is an apt description of the British worker from the dawn of the Poor Laws through the rise of Fabian socialism.

Marx failed so miserably in his effort to put an end to the confusion he found in Wakefield that America, where the population retains still its relative immunity to "expropriation," and where wealth is still broadly dispersed, is considered to be the ultimate capitalism on precisely those grounds, by Marxists first of all.

Of course the argument will be made that America has changed since Marx wrote. This is certainly true. It has changed so utterly as to make comparison futile, in all but the broadest terms. Broadly speaking, then, it remains true that the American economy is character-ized in a high degree by the accumulation of "self-earned private property," sufficient in most cases to allow people to change jobs, acquire skills, strike, move into a better labor market, become self-employed. This minimal independence, an effect of relatively high pay, allows those who work for wages to demand relatively high pay. Marx saw capitalism growing on the tomb of the "colonial" economy, and it may yet. When it does, the signs will be a marked decline in independence, mobility, and standard of living—the things we Ameri-cans associate with "capitalism."

It will be objected that we have already suffered a decline in these things—by those who know nothing at all about America in the nineteenth century. It is hardly likely that a reader as voracious as Marx was would not have come across accounts of crises, bankruptcies, and so on in the United States. They were neither rare nor unremarked, nor was the poverty that attended them.

Clearly, however, there was a degree of difference which amounted to a qualitative difference in Marx's view. Marx, repeating Wakefield's comparison of what Marx wishes to show are antithetical systems, says that Wakefield "depicts the mass of the American people as well-to-do, independent, enterprising and compara-tively cultured, whilst 'the English agricultural labourer is a miserable wretch, a pauper . . . In what country, except North America and some new colonies, do the wages of free labour employed in agriculture, much exceed a bare subsistence for the labourer? ... Undoubtedly, farm-horses in England, being a valuable property, are better fed than English peasants.' " Marx argues that the margin above subsistence which allows for culture and independence—and he is speaking here of farm laborers and "peasants" as such, hardly a prosperous group—makes an absolute difference in the character of the society. This conclusion is prepared by the eight hundred or so pages that precede it: a rehearsing, in large part, of the parliamentary reports, which document again and again expropriation and immiseration of an order so extreme that we might measure the relative happiness of our history by the difficulty we have believing it could be true. Capitalism, as Marx used the word, is very well described as growing on a tomb. I mention Sellafield here, how it thrives on poverty and unemployment, and never alleviates it, despite the great returns it boasts of bringing home to Britain. Marx, paraphrasing Wakefield, says, "But, never mind, national wealth is, once again, by its very nature, identical with misery of the people."

Americans are surprised when they learn that they are mentioned in *Capital* at all. They think of themselves, at best, as rather like Virgil and Plato, excluded from the scheme of salvation by accident of birth among the pagans. At the same time, oddly, they have appropriated the myth entirely to themselves, to the extent that, rather than looking into *Capital* to find out what Marx had in mind, they use themselves as the definition of capitalism and judge Marx as they feel kindly or unkindly disposed toward themselves.

Americans are startled to learn that Karl Marx was

employed as a foreign correspondent by the *New York Daily Tribune,* from 1853 to 1861, during the years in which he was preparing *Capital.* His *Tribune* articles, of which there are a great many, describe European and British affairs, the possible entry of the British government into the Civil War on the side of the South being an especially urgent interest of Marx. At the same time, though for a shorter period, Marx wrote about American affairs for *Die Presse,* a newspaper in Vienna. The articles were based on wide reading in American periodicals. Indeed, Marx said a great deal to Americans, and about them. He was in almost every instance more than generous, doing his American readers a courtesy to which they are now little accustomed, by addressing them always as humane and intelligent people.

Marx submitted articles to Horace Greeley, whose *Tribune,* published across the country in daily, semi-weekly, and weekly editions, was the largest newspaper in America, important to the development of abolitionist feeling and to the founding of the Republican Party. Marx's columns were often published in all three editions, sometimes unsigned as editorials. Abraham Lincoln is known to have perused the *Tribune* in the years before his presidential candidacy, while he was developing his political position. If he looked to the *Tribune* in 1861 for news of British intentions with regard to the war (preventing Britain from entering the war on the side of the South was his most pressing foreign policy problem), then Lincoln read Marx as President. Given the importance of the *Tribune,* and the long list of its distinguished contributors, and Horace Greeley's ties with Northern literary society, it is reasonable to assume that Emerson and Whitman read him, too. Harriet Beecher Stowe wrote a short life of Lincoln that appears to echo at least one of Marx's essays. Marx had written about Mrs. Stowe, so it seems all the more probable that she was aware of him. There was so much shared language among Marx and his American contemporaries that influence is not readily established. Greeley delivered attacks on wage slavery, and Marx described the condition of the British workers as slavery and called for their emancipation. An early essay of Friedrich Engels demonstrated the practicability of communist societies with accounts of American Shaker settlements, communal farms in the Midwest, and with mention of a community in Northampton, Massachusetts. Every example was drawn from the United States.

More generally, as Marx and the abolitionists both insist, the circumstances of workers of the time made slavery, actual as well as virtual, a live prospect for them. At least one Southern writer, George Fitzhugh, proposed ending the injustice of black slavery by enslaving whites as well. The revered Thomas Carlyle had called for the enslavement of workers in Britain, and lesser lights kept the idea alive, down to the time of the Fabians.

Abraham Lincoln's origins were of a kind to have made the enfranchisement of his own family fairly recent. Considering the legal constraints that had always burdened the poor, and the fragility of such rights as they had gained, there is a special resonance in his saying, "He who would be no slave must consent to have no slave." The boundaries human bondage threatened to overleap were racial as well as geographic. Conversely, its suppression had vastly more than national and humanitarian implications. And that is why, calling the Civil War a "working man's revolution," Harriet Beecher Stowe wrote:

The revolution through which the American nation has been passing was not a mere local convulsion. It was a war for a principle which concerns all mankind. It was the war for the rights of the working class of society as against the usurpation of privileged aristocracies. You can make nothing else of it . . . The poor labourers of Birmingham and Manchester, the poor silk weavers of Lyons, to whom our conflict has been present starvation and lingering death, have stood bravely *for* us. No sophistries could blind or deceive *them*; they knew that *our* cause was *their* cause, and they suffered their part heroically, as if fighting by our side, because they knew that our victory was to be their victory. On the other side, all aristocrats and holders of exclusive privileges have felt the instinct of opposition, and the sympathy with a struggling aristocracy, for they, too, felt that our victory would be their doom . . .

In the 1867 preface to *Capital* Karl Marx wrote:

As in the 18th century, the American war of independence sounded the tocsin for the European middle-class, so in the 19th century, the American civil war sounded it for the European working-class.

I am haunted by the sense that a changeling has been put in the cradle of American culture. Adam Smith, the supposed capitalist, whose influence among us is notorious, developed an economic system in which prison, the poorhouse, and starvation have no role, and in which the flourishing of the people (a term he prefers to "the poor") is the desired end. Compare the Fabians, those most sedulous strainers of mercy. Why are Smith's proposals for public projects to enhance the economy, taxes that weigh less heavily on the poor than the rich, and education to alleviate the effects of industrial work, called capitalist, while subsidies of the cost of labor, and visits of inspection to the homes of the poor to assure that their destitution was perfect before they were relieved—that women had sold their wedding rings, for example—are called socialist? Why do the Land Grant Act, the Homestead Act, and the G.I. Bill, three distributions of wealth to the public on a scale never contemplated in Britain, have no status among political events, when the dreary traffic in pittances institutionalized as the British Welfare State is hailed as the advance of socialism?

We must find a political and moral clarity that will enable us to address the starkest problems of survival, if the world is to have any hope. For a long time now, socialists have claimed an exceptional interest in the well-being of the generality of people, a special inclination to humanize collective life. But the history of socialism is disheartening. It is too strongly associated with repression, and these ties are too casually dismissed, for socialism to be conceded the special virtues it claims for itself. Plutonium manufacture and radioactive waste dumping are enterprises of the British government, and as good a proof as one could wish that government ownership in itself means nothing. The pattern identified by Adam Smith and Karl Marx, the accumulation of capital through the destruction of wealth, is fully present in Sellafield. British socialism has always been no more than the left hand washing the right, and yet for years it has compelled the admiration of American socialists, who can find nothing in their own tradition to compare to it for moral grandeur.

The mainstream political tradition in America is represented insistently now as unrelievedly "capitalist," whatever Marx might have said about it, and as compromised, grubbing, and mean-spirited because of the supposed relative prevalence of "private property"— whatever Marx may have said about that. On both the right and the left, capitalism, not democracy, is represented as the basis of our institutions. If Sellafield is sometime sold to private owners, as the government has long intended that it should be, then overnight it will become a classic capitalist enterprise by Marx's definition.

There is a third option, however, described by both Smith and Marx, and, as luck would have it, indigenous to America, of a society based upon individual autonomy, to be achieved through policies of government that by act or omission enhance the specific, tangible, material well-being of individual people, by creating or protecting conditions of life that enhance vigor and morale. These include education, fair wages, wholesome food and water, and reasonable hope for one's children. These things correspond in a general way with what Americans consider to be "Western values," yet they never have described, and do not now describe, the condition of life of ordinary British people. To the inevitable reaction, that people do not miss what they have never had, that the austerity of their lives has spared them the corruptions of materialism, that legal protections are needed only where society is a war of each against all, that there is the dole to assure them security from cradle to grave, however tedious the passage, or however swift, the reply must be that the history and present condition of ordinary British people make it clear enough how they have been used and in what spirit, by capitalists and by socialists, in tacit or declared collaboration. The best American political impulses occur outside this sham opposition. They need to be rediscovered as valuable impulses. Certainly we need to rediscover the complexity of our own political history, which deserves vastly better than to be seized upon by capitalists or dismissed by socialists.

When Abraham Lincoln said of a hypothetical black woman that "in her natural right to eat the bread she earns with her own hands . . . she is my equal, and the equal of all others," he expressed an economic proposition which is by no means commonplace or inevitable. Lincoln based the woman's rights on what she earned, not what she needed, a departure from "subsistence theory" and an implicit acknowledgment that labor creates value—that is, a margin between the cost of the worker's subsistence and the amount of wealth she creates—and that she has a right to share in this overplus. One learns from Adam Smith, Thomas Carlyle, E. G. Wakefield, and others that subsistence was assured to slaves as it was not to free workers. In Britain before the Second World War, employers still felt day laborers' arms before they hired them, so that men who were frail or malnourished could be turned away. Under ordinary circumstances slaves would have had as much as economic theory, up to the time of Beveridge, promised or allowed fully employed working people in Britain—enough to maintain them in a condition of physical efficiency. Lincoln made the case, successfully, that in justice more was due anyone. If he had used Marx's language, he would have declared the right to "self-earned private property." Against a history in which vulnerability triggered the crudest abuse—the history of the British poor, into which Africans were swept up fairly late—so modest a statement as Lincoln's sounds like beatitude.

The most difficult struggle of our civilization has been to find the means to create autonomy for ordinary lives, so that they might not be plundered or disposed of according to the whims of more powerful people. Ideas like civil rights and personal liberty come directly from this struggle, which is not terribly well advanced at best, and which is untried, failed, or abandoned in most of the world.

So very much depends on a poor man's wage. At present there is a Youth Training Scheme in Britain to absorb the energies of unemployed school-leavers. Industries are encouraged to take on trainees in place of regular hiring. These youths are paid by the government at wages that about equal the dole. In other words, the government donates to industry the free use of unskilled labor. Without reference to the wealth these young people produce, their subsistence is counted as welfare spending, and they are thought of as the beneficiaries of this arrangement, from which it is hoped they will learn the value of honest work.

People in their teens are historically the most coveted workers in the British economy. They are relatively healthy, and from the government's point of view, they are cheap, because they have no dependents and normally live with their families. This scheme merely reproduces the ancient pattern, severing work from pay, making the wage a charity, while reducing work to an escape from the opprobrium of idleness.

Britain invests more money abroad per capita than any other country, and invested more in absolute terms, until Japan surpassed it in this decade. Those who control capital, whether banks or industries or the government itself, have always had the means to punish or starve policies they disapprove of, or to crown with success policies friendlier to their interests, simply by leaving money at home or by siphoning it off to the United States, or to South Africa or elsewhere.

Mrs. Thatcher was described in an essay by Bernard D. Nossiter in *The New York Times* (June 15, 1988)* as arguing to a church assembly "that abundance, the rich, were blessed while poverty, the poor, were not, and Creation proves it."

She has her reward. Britain is experiencing economic growth, of the hectic, selective, up-market kind which does not threaten to drive upward the cost of industrial labor or the demand for social services. British economics is a game of keep-away. Whence all the jiggling of statistics—it is easy to get a big percentage increase from a very small base, as in calculating wages and pensions, and it is easy to take away with one hand what is given with the other, to raise wages a little and cut benefits more, and it is easy to increase rates of saving and contribution to private pension plans by reducing benefits for the elderly or cutting back on the administration needed to deliver them or adding to the obstacles involved in obtaining them or threatening to phase them out altogether, as the Thatcher government has done.

---

* Bernard Nossiter, *The New York Times*, June 15, 1988, p. A31.

Ralf Dahrendorf, in his book *On Britain*, quotes respectfully as follows from a book titled *Equality*, co-authored by Keith Joseph, an important political figure in the Thatcher government: "Ultimately the capacity of any society to look after its poor is dependent on the total amount of its wealth, however distributed." One might object that the way in which wealth is distributed determines, in a society, how numerous "its poor" will be. To distribute wealth away from employed people, as the British do, creates poverty, which must be looked after, perpetuating the ancient relation of those who work to those who employ, which has analogues, or cousins, in slavery and forced labor.

In the 1880s a philanthropic businessman named Charles Booth launched on a survey of the poor of London, with the purpose of refuting socialist assertions that a million Londoners, one fourth of the population, lived in deep poverty. He continued his project for seventeen years, in the course of it concluding that the extent of poverty had in fact been understated. His work, titled *Life and Labour of the People in London,* had a considerable significance for modern British socialism, in part because Beatrice Webb, the indefatigable propagandist and guiding spirit of Fabianism, cut her teeth in assisting these researches, and in part because Booth's work stands in the direct line of descent of British social and economic thought, specifically identifying the institutions and functions of Poor Law as socialism. Booth and the Webbs were associates of William Beveridge in his early career. Booth, like Beveridge, was a Liberal, in Britain the name of a party of bare-knuckled laissez-faire-ists, without any trace of what Americans call liberalism. This is considered an irony of British history, the importance of such people in the emergence of the Welfare State. It is no irony. Like all reformers before them, they undertook to design a good and harmonious state based on stabilizing the poor in a salubrious poverty.

Booth was a man of conscience, an advocate of reform, a visionary of the kind with which British history has long been replete. He dreamed of a day when "the streets of our Jerusalem may sing with joy."

Americans are often said to have a dream. It has never been clear to me just what I and my nation—Armenian dentists in San Diego, Norwegian Avon ladies in St. Paul, black nuns in New Orleans—fall to dreaming about so universally and persistently. The common wisdom is that we dream of personal success and prosperity. But we all know that those freshly arrived in this dreamy nation succeed and prosper at a rate that makes slouches of the rest of us, and that the naturalization of their children into our culture has no more striking feature than this tendency to drift downward from the heights of aspiration. So when these newcomers fall to dreaming along with the rest of us, what do they dream about? Personal sanctity, perfect crime, beautiful lovers, the admiration of their family, a house in the country, physical persons that mark them as vigorous, alert, temperate, and self-sufficient. It seems to me that people's dreams lie along the grain of their expectations, in general, so that chefs dream of owning restaurants, rabbis of being highly esteemed rabbis—in other words, that in America goal-oriented dreaming is as diverse as individual circumstance.

There is, however, a British dream—though I have

never seen the phrase anywhere before. It is an anxiety dream, of the lean kine eating the fat. Joseph in the Old Testament is summoned to interpret a dream of Pharaoh's, that seven fat cattle are eaten by seven lean ones, and that seven good ears of corn are eaten by seven poor ones. In the midst of plenty, Pharaoh has had a dream of famine. He assigns Joseph to store grain for seven years, in preparation for seven bad years, and disaster is averted. A glance at the world will assure one instantly that the dreams of cultures are much more typically urgent and recurrent anxiety dreams than the sort of delusional daydreaming Americans are said to be prone to.

In the British version of Pharaoh's dream, the lean kine are the poorest members of the population, who have half devoured the class just above them. The British, a practical race, and one whose prophets have warned them of the exponential increase in the numbers of lean kine, so that they have not enjoyed the assurance given to Pharaoh that his troubles would sometime end, have taken practical steps—in the thick of the dream, of course, from which they have never wakened. It was in his prosperity that this warning came to Pharaoh, and it has been through the centuries of British prosperity that their dream has haunted them. And here is the strange part, the place where British experience departs unequivocally from its biblical analogue. The very thrift and saving required to keep disaster at bay actually increased the numbers of lean kine and their voracity and—if I do not overstep the strange metaphor unpardonably—their rage. The alternative, as Malthus and others were quick to point out, was to feed the creatures, thereby accelerating their increase in numbers, without finally diminishing their voracity. Wealth does not deserve the name which must constantly be guarded and worried over, and the British, however rich, have always felt poor. But with Charles Booth we have arrived at that cusp, that threshold moment, when capitalist Britain is about to emerge as socialist Britain, and the nightmare is finally about to end. If we pause here we can witness this epochal change. It requires careful watching, for there is as little change to the outward eye as there is in bread after consecration.

Though Booth himself was no socialist, he proposed that socialist institutions—prisons, poorhouses—might act as a catchment for those elements which cannot function in an individualist culture. In what particular this would depart from the state of things then prevailing I cannot tell.

Friedrich Engels's study of the conditions of the working class in Manchester tends to describe streets and quarters—how they were built, how populated, what sanitary conditions prevailed in them—or to describe categories of workers and factory conditions. Booth turned much of his attention to families, which he describes in terms of their cleanliness, thrift, and sobriety, the presence or absence of which correlates with their class ranking on a scale of his own devising. Just as Engels's method reflects his belief that the working class were victims of their circumstances, Booth's reflects the more usual assumption that moral failure is the great cause of distress, with the inevitable corollary that poverty, if it is done right, is entirely consistent with happiness and well-being. Thus, for Booth, the cause of misery is found in the miserable.

Charles Booth did not, however, foresee abandoning the feckless to their misery—they were, after all, a drain on the economy: "Ill-paid and half-starved as they are, they consume or waste or have expended on them more wealth than they create."

His solution will sound familiar. Class A, his designation for "the lowest class of occasional labourers, loafers, and semi-criminals," could be "gradually harried out of existence." The more fortunate Class B, the "ill-paid and half-starved," are still, in Booth's elegant phrase, *du trop,* a phrase that should surely be translated as "redundant." His scheme for eliminating this class is gentler than harrying them out of existence, though not by much. In his view, "if they were ruled out we should be much better off than we are now; and if this class were under State tutelage—say at once under State slavery— the balance-sheet would be more favorable to the community." Like Carlyle, he considers slavery a humane reform, but unreachable. For him the difficulty lies "in the question of individual liberty, for it is as freemen, and *not* as slaves, that we must deal with them. The only form compulsion could assume would be that of making life otherwise impossible; an enforcement of the standard of life which would oblige every one of us to accept the relief of the State in the manner prescribed by the State, unless we were able and willing to conform to this standard. The life offered would not be attractive."

Let us pause here to note certain features of this plan. First, the social amelioration it has as its object would "rule out" a class of people who supposedly represent a drag on the economy of a nation which was at that time the richest in the world, and in the history of the world. The shared characteristic that defines them as a class according to Booth's system is their extreme poverty, which, as he freely grants, can be brought on by illness or accident. Nevertheless, they are "as freemen" to be compelled to "accept a regulated life," in "industrial groups, planted wherever building materials were cheap." Granting that these groups would not be economically competitive, "it would be merely that the State, having these people on its hands, obtained whatever value it could out of their work. They would become servants of the State." Accounts should be kept of the value produced by each family, expressed as deficiency, of course, since the enterprise would not be profitable. "It would, moreover, be necessary to set a limit to the current deficiency submitted to by the State, and when the account of any family reached this point to move them on to the poorhouse where they would live as a family no longer. The Socialistic side of life as it is includes the poorhouse and the prison, and the whole system, as I conceive it, would provide within itself motives in favour of prudence, and a sufficient pressure to stimulate industry."

The thing most striking about this proposal is that there is nothing new in it. It criminalizes poverty, making poverty entail penal servitude. Like the fourteenth-century Ordinance of Labourers, it assumes that those who do not work continuously are, even in their misery, an expense to the state, and that the state has a right to make the balance sheet more favorable by seizing upon their labor. One wonders what form coercion would take if it were not limited by the "freedom" of its objects. The use of separation of family members as leverage on their behavior—specifically, in extracting economic value—was a feature of the New Poor Law of 1834, notorious and widely practiced. Even infants and mothers were separated. Booth is simply pointing out that the poorhouse, a part of "the Socialistic side of life as it is," would still have terrors for his Class B "sufficient to stimulate industry." Booth's scheme would merely extend the sanctions against pauperism to include those who had eluded that status by scrounging and suffering, through religious or private charity, or through the openhandedness of the more fortunate poor. Like the old vagabonds, they would be punished in anticipation of transgression, for "to bring Class B under State regulation would be to control the springs of pauperism." Booth offers in support of his proposed reforms to eliminate the non-productive classes the observation that "the Socialists think it can be done by self-devotion on the part of the capable, and a final sternness which shall enforce obedience by the threat of starvation."

One can only envy the clean conscience a society must enjoy that so infallibly locates the sources of social problems in those who suffer them. If poverty is a transgression, it is one of which British opinion makers have always been, to an extraordinary degree, innocent—they need hardly fear disadvantage if they should sometime be measured as they mete. Yet to accuse them of a lack of fellow feeling with the poor is clearly wrong. They assume in the poor emotions they find most estimable in themselves, notably love of country and love of family. We are assured of this by their confident use of the threats of forced emigration, and separation of families, in regulating the poor.

There was, as I have said, a Minerva fully armed, growing within the squamous limits of Booth's undertaking, by which I mean, of course, Beatrice Webb, at that time still the parthenic Beatrice Potter, who, with her future husband, Sydney, would become the dominant spirit in the Fabian Society, the most characteristic and influential school of English socialism. She was the niece of Booth's wife, the daughter of a wealthy manufacturing family, and her curious lot had been to have her education entirely at the hands of Herbert Spencer, exponent of Social Darwinism.

One is supposed to admire the Fabians, just as one is supposed to admire English socialism. The entire Fabian *corpus* reads as if printed in embalmer's fluid. And, as with Marxism or Structuralism, the tediousness of the prose acts as a defense of the ideas expressed in it, since none but the devout can endure reading it.

Beatrice Webb wrote the chapter in Booth's study devoted to the Jews of London. It is uncharacteristically lucid and evocative, and—by the standards of the whole work—eminently fair. Jews were clean, sober, literate, frugal, hardworking, debt-paying, and, relative to the larger community at least, independent. They were imbued with every grace Booth's study was meant to measure, and Webb allows as much.

The late nineteenth century was a period of Jewish immigration to London from Russia and Poland. The Jewish community of London established a Board of Guardians to provide for their poor, who were of course very numerous. Not surprisingly, considering the mentality at work, this arrangement was attacked in the London press as creating Jewish pauperism—since to receive assistance has long had, in the British mind, a disastrous and nearly irreversible impact on the human character. This "demoralization" will obsess Beatrice Webb. That it is an authentic and important phenomenon she never seems to doubt, and no wonder, since it has been a part of the prevailing view of man and society in Britain over centuries. So the Jews, in the seemingly unexceptionable course of taking responsibility for their own poor, were seen to be creating a "pauperized," that is, a demoralized or degraded, population.

Webb defends them, however. She says it is unfair to describe the providing of free funerals as proof of pauperization, because of the "peculiar solemnity of mourning and funeral rites among Jews, and the direct and indirect costliness" of them. Now, as it happens, nothing seems to have mattered more to the Gentile or Christian Londoner than his own or a relative's funeral. Working-class people, even young children, joined "death clubs" to pay the cost of their own funerals which, even after the Second World War, in William Beveridge's terms, behaved as a necessity. That is, people would keep up their payments through hell and high water. Booth reports that poor people felt it a point of honor to bury someone in the best style his savings would permit. In the previous century the bodies of deceased paupers had been dumped in open pits. This no doubt accounted for the passion of the poor for funerals, while establishing in the official mind a standard of economy never again to be attained.

One of the lesser Fabians will suggest that the cost of working-class funerals, to be assumed by the state, should then be lowered by introducing mass production of coffin handles. This is an early and characteristic example of the eagerness of British socialists to use state services to control and depress working-class consumption.

Again remarkably, Webb defends Jewish charitable practices on the grounds that they consist largely of "business capital of one kind or another, enabling the recipients to raise themselves permanently from the ranks of those who depend on charity for subsistence." She does not conclude that the phenomenon of pauperism might be perpetuated among other Londoners by the meager and abusive charity they enjoy at the hands of their Christian nation.

Yet among the poor themselves there is a generosity so considerable as to defeat these philanthropists' hopes of rooting out pauperism by controlling charity. Webb frets that, while the Jews can keep good records of those they relieve, "owing to the fact that our indigent parasites are to a great extent maintained by the silent aid of the class immediately above them, we can by no possible means arrive at an approximate estimate of the number of persons in our midst who depend on charitable assistance for their livelihood." That these "parasites" avoid the legal status of pauper merely frustrates her. While they exist they offend.

After all she has said in their favor, the future Beatrice Webb cannot finally approve of the London Jews and their philanthropy. Despite its apparent success, it is socially destructive, because "if we help a man to exist without work, we demoralize the individual and encourage the growth of a parasitic or pauper class. If, on the other hand, we raise the recipient permanently from the condition of penury, and enable him to begin again his struggle for existence, we save him at the cost of all who compete with him . . ." The impact of the Jew is therefore finally destructive, "for the reader will have already perceived that the immigrant Jew, though possessed of many first-class virtues, is deficient in that highest and latest development of human sentiment—social morality." Inured to deep poverty, he will underbid other workers without the restraints of "class loyalty and trade integrity," thereby lowering wages—"without pride, without preference, without interests outside the struggle for the existence and welfare of the individual and the family." These are the same people who, a few pages before, were faulted for their ready and substantial philanthropy.

Having documented exceptional independence and prosperity, she concludes sepulchrally, "In the Jewish East End trades we may watch the prophetic deduction of the Hebrew economist [she means Ricardo] fulfilled—in a perpetually recurring bare subsistence wage for the great majority of manual workers"—writing here for all the world as if Ricardo's prophetic deduction were not fulfilled in every corner of Britain.

Consistency is not the point. The point is to discredit the phenomena of self-help and social mobility, to find positive harm in them despite any apparent good. That Karl Marx, an immigrant Jew in London, should have attached such value to these things in his final chapter of *Capital* seems natural enough in light of the community Beatrice Webb describes here. It is her view that has carried the day. "Marxists" cleave to her Ricardian definition of "class loyalty" as if it were an article of the true faith.

The rigors of the system Booth proposed, to end poverty in Britain by harrying the poorest out of existence and regimenting the class above them, are entirely compatible with socialist thinking as it developed in Britain and which he by no means unfairly associated with the poorhouse and the prison. In an essay included in *Fabian Essays in Socialism,* edited by George Bernard Shaw and published in 1889, Annie Besant describes thus the mechanisms that will encourage workers to fulfill their duties when socialism is achieved:

At first, discharge would mean being flung back into the whirlpool of competition, a fate not lightly to be challenged. Later, as the private enterprises succumbed to the competition of the Commune, it would mean almost hopelessness of attaining a livelihood. When social reorganization is complete, it would mean absolute starvation. And as the starvation would be deliberately incurred and voluntarily undergone, it would meet with no sympathy and no relief.

In other words, the new order would achieve that elusive object of all British reforms affecting the lives of workers, the isolation and elimination of the unworthy poor, worthiness, as always, determined by economic productivity as they chose to define it.

Competition for work cheapens the cost of labor and prevents it from rising in value when demand for workers increases. So the unemployed contributed, after their fashion, a fact grudgingly acknowledged in whatever tolerance they enjoyed. It is not irrelevant to my larger purposes to point out here how little awe there is for human life among these philanthropists or in the tradition they inherit. Human life is weighed routinely in the scales of simple commerce.

Shaw himself, decades later, fulsomely praised the head of Stalin's secret police, whose picture he kept on his mantel, for having shot an inefficient railroad worker. Shaw expounded on the merits of extermination as a feature of social policy during the formative stages of Russian socialism and German National Socialism, visiting Hitler and Stalin with praise and advice and the prestige of his Nobel-decorated person. For some reason his and the Webbs' enthusiasm for authoritarian movements has always been interpreted as a misreading by them of the nature of these movements. Yet it would not be difficult to defend Shaw's statement that Stalin was a good Fabian, his Soviet labor camps being item one. In Shaw's introduction to his late play, *On the Rocks,* his enthusiasm for extermination actually recommends the policies of Stalin and the National Socialists to him. For some reason, Shaw has emerged from these involvements reputation intact. He is an acerbic wit, we chortle at his outrageousness. In terms of his own tradition, his views on this issue are not exceptional. In 1839 Thomas Carlyle wrote, "The time has come when the Irish population must either be improved a little, or else exterminated."

It is difficult to absorb the fact that definitions of goodness, justice, and right vary wildly, especially difficult since we think and speak as though the "West" had reached profound consensus about these things through a long evolution, and was now unanimous in its loyalty to certain values—which were, it is true, swept away in a sort of mud slide of unhealthy enthusiasms only fifty years ago, an event that remains totally inexplicable so long as we insist that "Western values" include a steady faith in the importance of individual freedom and human life.

Because Americans believe that there are such things as "Western values" and that these are identical with their own best impulses, they cannot recognize the real content of the ideas carried along in history, though these ideas have blossomed in events and institutions such as the poorhouse, slavery, the penal colonies in Australia and elsewhere, labor camps for social parasites, and the extermination of unvalued people. I place Sellafield among these manifestations, because it is comparable in nature and in scale.

With the establishment of the Welfare State words of consecration were spoken over truly ancient mechanisms for maximizing profit, stabilizing social relations, and at the same time clothing the state in the garments of benevolence yet again.

William Beveridge, whose Plan for the Welfare State led to his being called the "father" of the new postwar Britain, merely amended Poor Law by absorbing the insurance systems that had been set up by working-class groups and labor unions into a system called National Insurance, taking the rates of benefit from those already established by the very poor to hedge against disaster. This policy was suggested as early as 1786, in *A Dissertation on the Poor Laws* by Joseph Townsend. Of course, the levels of benefit contrived by the necessitous to see them through their darkest hours were not generous. Benefits financed out of contributions or deductions from the pay of people so situated that they had to brace against the eventuality of a three-pound funeral were necessarily minimal. Beveridge's innovation, the conceit of calling the state the insurer, meant simply that taxes would take the place of insurance payments. Contribution would no longer be made at the earner's discretion. These changes infuriated private insurance companies, some of which had grown very great from this modest base.

The problem of poverty, as it was understood by William Beveridge, interpreting a survey by Seebohm

Rowntree of the poor in the city of York in 1901, was, needless to say, nothing so simple as lack of money. It was that income was unevenly distributed throughout a worker's life. He (using a male instance, as Rowntree did) was poor as a child too young to work, better off when he and his siblings could work to help their parents, poor when he had young children, better off when they were old enough to work, and poor again when he passed the age of employability, usually at about forty. It was Beveridge's idea—Rowntree protested against it—that the problem to be solved was the uneven distribution of money over these five stages of the worker's life. The "excess" of earning in his more prosperous periods should somehow be made to provide for him in the periods when his income fell below subsistence level. (Beveridge adopted Rowntree's definition of subsistence as being able to live without sustaining physical damage on an income devoted entirely to the purchase of necessities, very strictly defined.) Rowntree wrote a critique of the Beveridge Plan, arguing that many of those who fell below his standard of subsistence were fully employed families whose situation was the simple consequence of low pay. He urged a minimum wage—to no avail, of course.*

A tax system like Beveridge's, which withholds a substantial portion of even small incomes, to pay it back again as Child Benefit, Housing Allowance, unemployment compensation, medical care, old age pension, and so on, is a solution to what British reformers have always seen as the improvidence of the working class, who used their spare money, in the best times, on pleasures and comforts—tobacco, tea, clothing, and drink—and, as

---

* "Poverty and the Beveridge Plan," B. Seebohm Rowntree, *The Fortnightly*, February 1943, pp. 73–80.

Beveridge noted, on movies, where, as Orwell noted, they often went to stay warm. This is the sort of excess to be squeezed out by "redistribution of income," which in British parlance means the reapportioning of income through a single worker's life, not from the more to the less well off.

Taxation regularized the insurance system, at least to the extent that participation was made universal among employed people at a certain wage level. What it omitted to do was to guarantee a return on the investment at any specific rate. The British government turns a profit on the National Insurance System, which goes into the treasury. So those who pay into National Insurance are taxed at a rate that subsidizes other activities of government rather than enhancing services or lowering rates of contribution. This merely reflects the larger fact that, in guaranteeing subsistence, Beveridge designed a system which would yield neither less nor more. Only against a history of severe deprivation could such an arrangement be called a Welfare State. Only a history of unlimited expropriation would make the Beveridge Plan read as a promise rather than a threat. Yet, a true descendant of the Poor Law, this system is treated as generosity so wanton as to have virtually destroyed the national character.

But the fact that there is in the system no guaranteed rate of return on contributions introduces the flexibility which allows the government to control the rate of real wages. In the nineteenth century, industrial employers could pay wages in kind. A worker might receive as pay a quantity of cloth of a very poor quality, and unsalable. This practice was one means—there were many—by which the employer could depress the cost of labor. The British government, by maintaining a level of service below the money value paid for the services, depresses the real wage of workers by selectively depressing the value of what the worker receives in lieu of money withheld in insurance contributions. The Welfare State is descended from the Company Store.

This assuming, by the British state, of the role of industrial employer in depressing the level of real wages is natural enough, since the state has always been identical with agriculture, banking, and mercantile and industrial interests. The idea that the British establishment predates materialism somehow (so much for Croesus), and is rendered free of mercenary considerations by sheer antiquity, is the sort of hokum that ought to make anyone feel for his wallet.

Without going into the details of arrangements that vary from case to case, I want to suggest that the nationalization of industries has also functioned as a refined form of expropriation. The National Health Service, for example, buys drugs from the major British drug companies, all of them private, on terms highly favorable to them. Its budget includes the cost of training all doctors, even those who go on to practice in the numerous private hospitals, which are run by private health insurance companies. So the state in effect subsidizes major private industries with money supposedly spent on public health care—health care, first among all those killing instances of largesse which are the boast and lament of contemporary Britain, just as the Poor Laws were their boast and lament in the centuries preceding.

As it happens, almost no one in the West spends as little on health care as the British,* despite the fact that

---

* See "Britons top lung, heart death list," *The Times*, January 8, 1987, p. 7.

they lead the world in death rates from heart disease and lung cancer, costly diseases, if they are cared for. All the noise is simply an especially dramatic manifestation of the princess-and-the-pea paradigm, where the British establishment takes a horrible bruising from an irritant no one else would feel. But there is a wisdom in these tribal rituals. All the lamenting over the burden of public expenditure gives the government a fine reputation for public service, at home and abroad, while it creates an atmosphere congenial to hospital closings— tolerant of, for example, twelve-month waiting lists for spinal injury patients.

One need not look too deeply into the economics of socialized medicine to account for the state of the National Health Service. The government has simply exercised control over consumption by pegging costs at 5 percent of gross domestic product and squeezing services to fit the budget. With an aging population, living at close quarters in deteriorating cities, and with a declining domestic economy—the base against which the budget is figured—there is a rise in need and a corresponding contraction in the resources for responding to it.

Mrs. Thatcher shuts down industries in the service of economic reform. I admit to finding her wisdom unsearchable. But the antecedents of her policies can be clearly seen in British history, when with bold strokes, implacable governors (call them lords or ministers or manufacturers) have swept away the livelihoods of great numbers of British people, in the service of some high object, like putting the Great back into Britain, which always involves enthusiastic obedience to economics' brazen law, the subsistence theory of wages. Supposedly, dead regions of modern

Britain will regenerate themselves. These disused populations will go forth and strive and innovate—a feat all the more glorious since it will be done without resources or training. I think it is more probable that they will become poorer and poorer, that the British economy will concentrate itself even more intensely on white-collar, middle-class service industries such as banking and insurance, in foreign investment, in toxic and radioactive waste disposal, in the chemical and drug industries, the weapons trade, and "invisibles." The old industries which labor seized upon as its domain are taken away as the commons were from its ancestors. The dole, by preventing, in theory and in general, the worst consequences of these revolutions from the top, makes such high-handedness possible. Poverty as a norm of working-class life has made them acceptable.

Welfare may be seen to be a matter of definition, having to do with values and expectations. We in America have not yet learned to congratulate ourselves for maintaining millions of unemployed people at the level of subsistence, and paying millions more who are employed, at wages that must be topped up to reach the standard of subsistence. One third of the British population is in poverty as the British define it.

Beveridge's plan, a national compulsory insurance system which assumed—more precisely, was contingent upon—full employment, incorporated union and "voluntary" schemes for covering births, deaths, injuries, illnesses, and temporary unemployment. It adopted also the long-debated system of family allowances, which added to family after-tax income with each child after the first.

The insurance element of the plan simply appropriated the shifts workers had made, to pool a part of their income to fend off the bleakest eventualities, appropriating at the same time the income of workers to cover the costs of the burden the state had assumed. Family allowance was distinctly less popular in its origins than national insurance. It was an attempt to nuance levels of income to conform to need. William Beveridge actually made a case for it first in 1919, when it was his responsibility to attempt to solve problems that beset the coal industry. The solution he hit upon was, I need hardly say, to lower wages.

He proposed to minimize hardship by lowering most the wages of those with fewest dependents. The union protested violently against the entire proposal. In those days there was talk of establishing a minimum wage in Britain. The family allowance system, by allotting pay on the basis of need, was contrary in principle to the idea that the laborer might unconditionally deserve a specified rate of pay. There is no minimum wage in Britain now; the family allowance system has prevailed. Workers, who are heavily taxed, have their money eked back to them—the payment is in fact made to wives—as a very modest subsidy calculated by the number of children, excluding the first child, for whom it is believed best that his parents be "responsible." Another among these grinding benefits is a mechanism whereby one may apply for some unreachable necessity, a pot or a blanket. In these frugal days, costs are to be "clawed back" (truly one of the great phrases) through withholding from pay or benefit. This extraordinary system is designed to keep real wages at the lowest possible level.

And what are the limits of the possible in this situation? When Eleanor Rathbone plumped for family allowances early in the century, she was encouraged by the discovery of recruiters for the Boer War that starveling men make unsatisfactory soldiers. Malnutrition was a threat to national security. Her view was that workers failed to maintain themselves at the level of "physical efficiency," not because they were poor, but because they made a bad job of being poor. Family allowance was to guide the expenditure of income, not add to it.

It is the distinctive achievement of British socialism to have attracted the energies of the clever to sorting out the details of the lives of the poor. It makes perfect sense, given the assumptions of the culture, that the solving of difficult social and economic problems should be the work of those with heads for that sort of thing.

From the beginning, however, the object was not to eliminate poverty but to make of poverty a less injurious condition—more precisely, to allow the commonwealth the economic advantages of poverty without its economic disadvantages. Alleviation of the conditions of the poor would create a more productive work force, in theory, though in fact British industry had thriven for more than a century on the labor of hungry and exhausted people, largely women and children, and had found little use for vigorous adult men, except in mining and shipbuilding. Industrialists had made a spectacular demonstration of the economic viability of existing conditions. Indeed, the decline in Britain's dominance of the world economy was usually laid to the *rise* in the living standards of the working class, even though the greatest emerging competitor, the United States, had the world's highest wages. World opinion might seem to set a lower limit to the standard of life a

wealthy power can establish for the mass of its people
and still enjoy a good name, but British experience
continues to prove otherwise. So the floating downward
of real wages has no obstacle or limit. This is truer now
that weapons have evolved which make armies redundant.

What living standards really are at one time or
another in Britain is a difficult question. Discoveries of
poverty always come as a great surprise—to the newspapers, at least. This indicates that the imagination of
the conditions of life among those who opine on such
subjects is consistently wrong. In the nineteenth century
an outbreak of cholera would produce a burst of
information about misery and crowding and unwhole-
some conditions. Parliament would inquire and report,
pestiferous slums would be razed, and quiet would settle
again over "public," that is, polite, consciousness, before,
alack, the poor who had lost their dwellings were
provided with others. Slum "clearance," like the pulling
down of villages that depopulated the English countryside, merely emptied an area of irksome people. Once
out of sight, they could be forgotten.

The Road to Wigan Pier, which George Orwell wrote in
the thirties, is a sort of song of innocence and experience
which shows with unusual clarity how two contradictory
apprehensions of working-class life coexist in one
"lower-upper-middle-class mind." Orwell has pene-
trated this life, not quite as a "visitor," one of the
philanthropic inquirers who entered the houses of the
poor to inspect for demoralization, but as something
very near akin. He uses the word "inspect" with irritat-
ing frequency, and he discovers demoralization. The
conditions Orwell describes—poor food, crowding, brutal and uncertain employment—are the staples of this

sort of writing. Orwell shares dirty food and a fetid bedroom. The inspiration to stay in a boardinghouse in order to observe his subjects intimately might have come from, for example, Charles Booth, who likewise claimed to have formed an affection for the classes in which he immersed himself, and who likewise reported that a particularly intense happiness was the good fortune of the typical working-class family.

Through much of the book Orwell particularizes his aversion to these people by describing his intimate observation of them. He finds them bitter or imbecile and uniformly evil-smelling. He remarks on the physical revulsion one of his caste feels for the lower orders, confessing that while he could not abide being touched by an English valet, he had not objected in Burma when his Asian "boy" dressed him. In any case, at the end of his bleak account of Wigan he describes his fondest memory, which is, inevitably, of a scene he did not see. Certainly like nothing at Wigan. It is a memory, not specific as to place or occasion or the people involved in it, of a working-class family gathered contentedly around a hearth: mother, father, son, daughter, dog— Mum serene over her sewing, Da in those tranquil throes of pleasure stimulated in working-class men by racing news. I think it must be to this scene that all such observers refer when they describe working-class life, which is always assumed to be good and comfortable, when it is posing no egregious problem of which the press or Parliament must take note.

Orwell's book demonstrates that even immediate experience cannot touch or disrupt this ideal, and equally that his idealization removes nothing of the stigma he attaches to the evil-smelling classes. Rather, he reinforces the aversions of his own class by allowing them the authority of an eyewitness. At the same time, his sentimental evocation of the special happiness of working-class life defines his hopes for them. He expressly dreads a future economic security that will rob the racing news of its fascination.

The present British government is pulling down the great edifice of British philanthropy. Margaret Thatcher is full of busy invention, for example, the poll tax, which will tax by the head, assessing everyone over eighteen equally, rather than taxing property. Taxation will be based on voting lists. The logic is very straightforward. People who live crowded in cheap flats are the primary consumers of social services, so they should be the ones to pay for them. The reform will induce responsibility, since people will be reluctant to vote for services if they must count the cost. The purity of cynicism attained in these reforms is oddly hilarious. The best of them so far has been the dream-of-home-ownership scam. The Thatcher years have seen, along with burgeoning poverty and raging unemployment, a vigorous increase in the numbers of homeowners. How has this miracle been accomplished? By raising rents, and at the same time offering mortgage payments lower than the new rents.* So poverty actually spurs this fancied embourgeoisement. Do these policymakers laugh over their work? Having bought a council house to avoid the cost of continuing to rent it, let us imagine that Mr. and Mrs. Homeowner suffer a setback—say, the loss of a job, or a cut in some welfare allowance, or

---

* See "British spending on housing 'lowest in world,' " Richard Thompson, *The Times*, November 20, 1984, p. 3.

a rise in the cost of living, or a rise in the interest rate, which is reflected in all British mortgage payments. Any of these contingencies is highly probable. Then Mr. and Mrs. Homeowner lose their dwelling, and it passes into other private hands. Creating all these private owners of erstwhile public housing is simply a way of destroying public housing. Foreclosure rates are the highest in history. Already, council houses in rural towns are being bought up by city dwellers as second homes. Those displaced from desirable housing will be crowded into undesirable housing, which demand will make expensive. The housing of the British working class is, historically, the crowning scandal in an appalling record of abuse. Public housing was the great postwar pledge that all that misery had ended.

In my eagerness to share my appreciation of the devilish wit of Conservative housing reform, I have skipped over certain implications of the new per capita tax which are well worth considering, including, of course, its impact on the finances of low-income home buyers. Margaret Thatcher is Nassau Senior back from the dead, and reliving that great period, after 1834, when, by enhancing the severity with which paupers were treated, he caused them to disappear by tens of thousands. In this milder age, a poll tax could work as well.* Since mere existence would imply a tax liability, people for whom a tax would be burdensome might tend to avoid drawing official attention to their existence, especially if they fell into arrears. They might not report crimes against them, or seek medical treatment, or sign up for the dole, or register to vote. A new

---

* See "Poll tax could 'shrink' population," Hugh Clayton, *The Times*, April 3, 1985, p. 2. See also "Young unemployed 'risk prison under poll tax,' " Colin Brown, *The Guardian*, March 26, 1986, p. 5.

buoyancy would come into the indicators of social well-being if the unfortunate classes kept themselves to themselves. School-leaving age will probably soon be lowered to fourteen—already only one child in five is in school after the age of sixteen—and schools are springing up which do not require attendance, an answer to the crowded classroom. "Redundant" British people have always been invited to disappear, and then they are found again by the pornographers of squalor, who stimulate cries for humane reform, the object of which is always to make the poorest disappear, by pulling down their slums, or encouraging them to emigrate, or punishing them to improve their character and encourage independence.

It is characteristic of the British official mind to take things to a certain point, and then, as it were, go blank. If only the worthy destitute should be helped, what should happen to the unworthy? How can a subsistence wage be calculated if workers have dependents? If you pull down a slum, where will the slum dwellers go? If you cut back health care for a poor and aging population, what will the consequences be? If you pump plutonium into the sea, will it return? Things, people, consequences disappear in Britain, into a deep reservoir of denial. They surface frequently, but not for long. The government is not observant or reflective, but invasive and peremptory, improvised, as if distracted from more important business. Yet, like an authentic modern government, it deals in the lives and safety of people, and enjoys the extraordinary powers conferred on governments by high technology. To be able to imagine actions unshadowed by their consequences is a source of enormous confidence, and great savings, but of neither wisdom nor moral seriousness.

The depth of British memory, as it can be seen in the recurrences of highly particular notions and obsessions, is at the same time remarkable. The poll tax exempts real estate from taxation. Similar logic was used at the beginning of the century to exclude landed property from taxation. The impulse to shelter wealth on the grounds that those who burden society should bear the cost of the burden they constitute—the rationale of the poorhouse—has persisted all these years intact. It is a systematic and principled rejection of the idea of community. The Fabians wrote approvingly about "creeping socialism," meaning institutions such as the post office and public sanitation, to which the mass of people enjoy access. State-supported education, that is, public education, is considered socialist as well. In other words, such inevitable enhancements of general welfare as derive from the evolution of civilized life in the last century and the first half of this one are considered the expression of a moral-political drift now being reversed. The Dartford Tunnel, a major conduit of traffic into London, is being leased to a consortium of banks, to be managed for their profit.

The bedrock British political assumption is that absolutely nothing belongs to the general public inalienably, by the logic of collective interest or by right. To understand why Britain has felt itself a Gulliver tied to the earth by innumerable threads of socialism, one must understand that public ownership of a bridge, a tunnel, or a river is for them a departure from the natural order of things.

*　　*　　*

William Beveridge went through a little interval of disfavor with the government when he submitted his report to the coalition dominated by Winston Churchill, a tight-fisted fellow even by British standards. The authorities first printed up an abstract of the report as if to distribute it to British troops in the field, then snatched it back at the last moment. Needless to say, the soldiers found ways to pilfer copies despite all, and the report was read avidly, with great excitement. I suspect this may have been no more than very deft marketing. Louis XIV, to make French peasants interested in planting potatoes, is said to have stationed armed guards around fields in which they were planted. The fields were in due course thoroughly plundered.

The British government rumbled and grumbled over Beveridge's proposals, while he toured an admiring America in triumph with a bride of mature years who wrote a book about the experience and its ironies. Meanwhile, in Britain, pressure in favor of his report grew.

There was apparently impassioned opposition, of exactly the kind the newest Poor Law should inspire. The Reverend W. R. Inge launched a Malthusian attack, calling it "a heavy bribe to the slum dwellers to have large families." In this view, "artificial dysgenic selection has never been carried so far" as in the Beveridge Report, "the most gigantic effort of blackmail ever made by a frightened government."* British soldiers dreaded the misery which they had left and into which they expected to return. Inge alludes darkly to the fact that

* "The Future of England," W. R. Inge, *The Fortnightly*, May 1943, p. 288.

slum dwellers have been discovered to be "a deadly danger in time of war." One comes across such remarks fairly often. Whether they refer to disorder at home or among the soldiers I have not discovered.

At any rate, in the midst of war, the dawn of a new order began to appear in the sky. In due course legislation based on Beveridge's plan was passed, supplemented with provisions for the National Health Service, education reforms, and industrial nationalizations. Interestingly, Beveridge himself lost election to Parliament. A Labour government was selected to preside over the *novus ordo seclorum,* on the strength of an overwhelming majority of the votes of the returning servicemen.

And the standard of living fell. The government used its new powers to lower wages, and continued to impose wartime rationing, in severer forms. Poor Law institutions were rechristened, hospitals and pensioners' homes purged of their bitter histories by a change of name. The National Health Service is still defended as a vast improvement over the horrors of the system which preceded it, also called the National Health Service, developed from the fact that poorhouses were increasingly the refuge of the destitute sick and old. Reports commissioned by the government, released in 1981 and 1987, indicate that class differences in illness and mortality have widened greatly and steadily from World War II to the present. Since, according to *The* (London) *Times,*[*] the "economically inactive," a group including "all illegitimate births and many of the permanently sick as well as many single parents," are not included in national statistics, the poor cities in the North, where health care is worst and unemployment exceeds 20

---

[*] "Does poverty equal poor health?" Richard Wilkinson, *The Times,* April 2, 1987, p. 2.

percent, would not figure appropriately in this measure of the success of the Welfare State. For it is the Welfare State, the high-water mark of British socialism, whose successes are to be measured in this decline in the relative well-being of the poor. Mrs. Thatcher's new dispensation can only exacerbate this trend, presumably, since poverty, unemployment, and radiation exposure, among other factors which correlate strongly with ill health, have all increased under her government.

I have no reason to believe that illegitimate births do not occur under the auspices of the Health Service, that the chronically ill do not die in hospitals, at least fairly often. How can the government not have information about these people? On what principle can it exclude such information? Does the midwife keep no record of delivering an unwed mother's child? Is National Health Insurance really insurance, in the sense that it covers only those who have paid for it? Then what becomes of the others? If there is a charity health system for the indigent, how can the government fail to have access to its records? If it has access to them, how can it rationalize their exclusion from health statistics?

I would suggest that we have here the modern incarnation of the unworthy poor, Booth's Class A. The Welfare State is made for the deserving, and desert is established by employment, as it has been for five hundred years. Are these "economically inactive" neonates the descendants of the wandering beggar women punished for burdening the parishes with their babies when Edward VI was King?

What would happen to infant mortality figures in America if official statistics excluded illegitimate births? What would official motives be in excluding such births? What would the government be expressing in terms of its social vision if it took no account of the condition of the most vulnerable? This British method of accounting is no recent innovation but established official practice.

Britain has only a shadow government. Its opposition is the shadow of a shadow, and fading. It is a government made not to shape reality but to conceal it. Yet the uncountenanced reality replicates itself like a compulsive gesture or an obsessive fantasy.

British records and social institutions are shielded by the laws that protect military secrets. There was a government attempt to suppress the 1987 report on inequalities in health that led to its being distributed from behind a guitar store, according to the press. Yet there is a profound respect for the conventions of secretiveness reflected in the willingness of distinguished men to compile and interpret information which conceals the conditions their work purports to describe.

Secrets are not merely kept, I think, but treasured. They give latitude to the old vice of punitive and abusive behavior, which lends piquancy to the great apparent seemliness of the people who preside over these same abuses. I have, over the years, gathered stories about the extraordinary exploitation of a boys' orphanage in Northern Ireland, and about an elderly woman horribly murdered, apparently by the police, because of her involvement in nuclear issues. Accounts of abusiveness and, especially, filth, in hospitals, prisons, insane asylums, and military training camps are very common, and simply too disgraceful to repeat. Anyone who is curious can go to the library.

In the nineteenth century the uncountenanced poor were called the "residuum." The same word was used to mean sewage. Scholars will note the powerful association of the socially rejected with filth. For example, British prisons have no toilets. Prisoners share densely crowded cells with a plastic bucket, which is emptied by them once each day. Some of these prisoners are debtors, of course, who have lived with such insult and nastiness since Britain first began its half millennium of misericordia.

That there should be a great secret, and a great denial; that the secret should involve filth and violence, in forms that are rarefied but at the same time quintessential; that there should be manufacture and world commerce and enormous profits involved, and a work force disciplined by poverty; all these things make Sellafield seem of a piece with its cultural setting. Finally, however, I am at a loss to describe the place it occupies in reality, wreathed as it is with distorted perceptions, with information pulled out of shape by the strategies of denial. I do not know the meaning of the violence the British government has done to its country and the world. I am sure no one could explain it to me. I think I am describing pathology.

In 1909 the quondam Fabian H. G. Wells published a novel titled *Tono-Bungay*, which anticipates the British nuclear enterprise in its most extraordinary aspect, the commercial importation of radioactive waste. Wells introduces the subject almost as an aside, yet with an eerie precision of detail. The stuff is called quap, ". . . the most radioactive stuff in the world. That's quap! It's a festering mass of earths and heavy metals, polonium, radium, ythorium, thorium, curium and new things, too . . . There they are, mucked up together in a sort of rotting sand. What it is, how it got made, I don't know . . . There it lies in two heaps, one small, one great, and the world for miles about it is blasted and scorched and dead. You can have it for the getting."

The quap lies along the coast, "an arena fringed with bone-white dead trees, a sight of the hard-blue sea line beyond the dazzling surf and a wide desolation of dirty shingle and mud, bleached and scarred." It is to be imported into England to make light-bulb filaments and gas mantles. These futurists should be listed too: British Nuclear Fuels now uses the radioactivity of gas mantles to make the point that radioactivity is a homely and familiar phenomenon. On two occasions British Telecom has disposed of tens of thousands of tritium-filled (therefore luminous) telephone dials as radio-active waste, and there are plans to bury a million more. False teeth and exit signs contain radioactive materials.* Since no law controls the use of radioactive materials in British products, no doubt other ingenious applications have been found for them. In Wells's fiction, the quap, which sickens the crew of the ship used to transport it, eats its way through the bottom and is lost in the sea.

Tono-Bungay is about British enterprise, a raw novelty in the early twentieth century, as it had been through the three or four centuries preceding. It eludes descrip-tion in the terms of traditional moral understanding, being, therefore, a vast field for opportunism and improvisation, as it had been for three or four centuries and as it is now. The obsessive bringing to bear of disapprobation upon "unprofitable" elements of the

---

* "Your daily dose of radiation," Geoffrey Lean, *The Observer*, July 13, 1986, p. 49; cf. also, "Chernobyl-style test in Snowdonia is off," Geoffrey Lean, Tony Heath, *The Observer*, March 6, 1988.

population has always implied an enormous freedom for those who float in the ether of profit.

The key to interpreting British behavior is always economic. Clearly H. G. Wells knew eighty years ago what consequences would follow from the accumulation of nuclear detritus along a coast. He wrote: "There is something—the only word that comes near it is *cancerous*—and that is not very near, about the whole of quap, something that creeps and lives as a disease lives by destroying . . . To my mind radioactivity is a real disease of matter. Moreover, it is a contagious disease. It spreads." Despite all that has happened since to confirm Wells's view of radioactivity, nevertheless "quap" has indeed been imported into England as part of a commercial venture.

Wells's anticipation simply demonstrates the fact that the British nuclear enterprise has never been innocent; that is, naïve. It is no more than a reprise of the sad old compulsions around which British social order has always turned. Such behavior would be modified, if not forbidden, if questions as to its wisdom or decency were ever raised in good faith. It is neither modified nor forbidden.

It is almost unimaginable that this industry could coexist with any lucid awareness of its implications, but this only means that our models for describing human behavior are fundamentally wrong. I suspect a better grasp of it awaits our recognition of major anomalies, that our conception of the fabric of motivation and causality must be warped into shapes that accommodate observable phenomena, with denial, dissociation, and atavism acknowledged as potent entities, like quarks and black holes. We should know by now the inadvisability of constructing a universe around local notions of reasonableness and plausibility.

PART TWO

Having come finally to my subject, Sellafield, I am forced to confront the epic scale of my narrative. My inability to invoke a suitable muse is really my only deficiency in treating this great subject. To the objection that I know very little about plutonium, I can reply that I know better than to pour it into the environment. On these grounds alone I can hope the British nuclear establishment will learn something from my work, so that I may repay them for the insights they have given me into the nature and prospects of mankind.

To the objection that I work largely from newspaper articles, I can reply that by the same means we learn most of what we take to be true; for example, that Margaret Thatcher is Prime Minister of Great Britain, and that her status has been arrived at by orderly means and carries with it significant prerogatives. It may well be that the moon landings were filmed in Arizona, and that the world's affairs are presided over by Freemasons, who stage elections and inaugurations only to mislead the rest of us. For all we really know, there is peace in Afghanistan and plenty in Ethiopia, and the

Irish Sea and the North Sea are of a most Edenlike purity.

Yet, granting the problems of knowledge, which are imposing, it is not generally considered prudent to discount entirely the information one finds in the press. Antinomians will wade into the sea I describe and delight as the warm ooze rises between their toes, and report that they have never been so refreshed. And any neoplasm that may afterward obtrude will be laid to a chemical additive in cereal packaging, or to the after-effects of Chernobyl, with perfect plausibility, though the only acquaintance most of us have with either of these phenomena is through articles in the newspapers. People believe selectively, and they are outraged selectively, so that any little area of informed or moral thinking tends to become a dot in a grand mosaic of pernicious nonsense.

It will become very clear that I do not invest great faith in any of my sources, no more in specialist publications than in those produced by self-styled champions of this unthinkably savaged planet. I have not written a "history" of Sellafield, because I doubt that it really has one, except insofar as a shopkeeper's ledger is a history. The supposed events that surround it seem purely epiphenomenal. To maunder on about leaks and fires is a bad joke in light of the fact that reactor cores are broken down and poured into the environment routinely and continuously. Such pother normalizes Sellafield, so that grave men can compare its "safety record" with those of other plants and industries.

Sellafield simply grows. Inquiries in 1976 and 1984, which enthralled the British press with calculations of peril and tales of malfeasance, coincided with two great expansions of the plant. I have read that in 1964

Sellafield was directed from defense toward commercial development by government policy. Commercial uses for radioactive materials are older than defense uses, however, and the British government are simply the last people in the world to arrive late at any chance to make money. So I am not sure that the plant has ever undergone any change at all, except in size. Its new facilities are being constructed at great expense (to the Germans and Japanese), to make it capable of extracting uranium and plutonium from new kinds of nuclear wastes.

The one thing always to be borne in mind is that, on the coast of Britain, wastes from a plutonium factory are poured into the environment every day. This is easily demonstrated from a little document, prepared by the British government Central Office of Information, printed by Her Majesty's Stationery Office, titled *Nuclear Energy in Britain.* I will quote first from the version published in 1976.

It comes as a disappointment to discover in such documents an important degree of impressionism in technical-sounding terms like "low-level" waste or "low-activity" waste. "Low-level" seems to mean no more than that the contaminant is intermixed with something else—it is on tissues, gloves, or overalls, or as the British use the term, it is in water, or in air. "Low-activity" wastes are relatively stable and persistent materials, like plutonium. So the effluents from Sellafield, insofar as plutonium is concerned, are low-level and low-activity; that is, widely distributed and with us forever. The pamphlet tells us, "Low-activity liquid wastes are normally discharged into public sewers, rivers, or coastal waters." It tells us, too, that "volatile fission products krypton, xenon and iodine," high-activity wastes, are in

"effluent streams" and are also "discharged, after treatment where required, to the atmosphere at heights
necessary to secure adequate atmospheric dilution,"
though in future they may have to be stored "until their
radioactivity has decayed." High-activity gases are
vented from smokestacks into that rainy climate in the
full flower of their brief lives, routinely, so that they in
effect combine the worst characteristics of volatility and
persistence. The pamphlet informs us that "the major
source of radioactive waste is the chemical reprocessing
of irradiated fuel elements," and looking to the future,
or perhaps merely speculating, it remarks that along
with other modifications "it should also be possible to
remove toxic elements such as plutonium from waste
streams."

This amounts to a fairly straightforward description
of the routine release of toxic and radioactive materials.
A revised version of the same publication, printed in
1981, says, "Low-level gaseous and liquid waste can be
dispersed directly to the environment where the dilution
of the waste is sufficient to avoid any significant risk to
the population." Notice what latitude the word "significant" allows to the expression of cultural values.
Further, "two main sources of low activity liquid effluent
arise at Windscale [that is, Sellafield]: water discharged
from the fuel storage ponds and waste streams arising in
the chemical plant. These are discharged into the Irish
Sea through a twin pipeline which extends 2 kilometres
(1.2 miles) offshore." An insight into the meaning of
"low-level" is provided later in the same paragraph:
"Low-level radioactive waste is also disposed of in
concrete-lined steel containers at sea, in the deep North
Atlantic Ocean, some hundreds of miles from land."
This is done in accordance with the provisions of two

agreements with truly wonderful names, the London Convention on the Prevention of Marine Pollution by Dumping of Wastes and Other Matter (1972) and the Multilateral Consultation and Surveillance Mechanism for the Sea Dumping of Radioactive Waste. According to the report, international inspectors come along to watch the barrels go over the side. In any case, low-level waste is the kind of thing one encloses in concrete and drops into the remote depths of the Atlantic, when one is not dispersing it directly into the environment.

These publications of Her Majesty's government document the essential fact, which is never disputed, that Sellafield pours plutonium and other radioactive substances into the sea and air. A booklet published in 1981 for the Commission of the European Communities, edited by J. R. Grover of the nuclear research center at Harwell, U.K., titled *Management of Plutonium Contaminated Waste*, makes a series of startling assertions about plutonium oxide which are apparently meant to justify the practice: that "its density is nearly that of lead which reduces strongly the possibility of it blowing over long distances"; that "it cannot be assimilated by plants or transmitted via biological routes"; that "in water its solubility is virtually zero. Therefore it cannot be transported by water."

In the first place, plutonium oxide forms extremely fine particles that become suspended in air, as the pamphlet itself acknowledges when it says of plutonium that "in a dusty form it is pyrophoric"—capable of igniting spontaneously in air. That this risk is compared to the danger of explosions in "other dusty operations" in industry merely restates the fact that plutonium oxide tends to become highly particulate. The pamphlet also describes dispersion of plutonium-contaminated gases through smokestacks, a practice which surely assumes that plutonium can be carried by the wind, since wide dispersion is supposed to render it harmless. The booklet actually defines gaseous plutonium-contaminated waste as "just the effluent air from plutonium process areas," an apparent acknowledgment that plutonium oxide is readily airborne.

As to the transmission of plutonium "via biological routes," plutonium concentrates in the liver, kidneys, and bone marrow, according to other authorities. This is to say that it passes into the food chain—into black pudding and kidney pie, for example.

The suggestion that plutonium cannot be transported by water because it does not dissolve in water suggests to me that the writer is not highly observant. Anyone who has hosed down a sidewalk is in a position to enlighten him.

The pamphlet goes on to say: "(a) There is no known chemical toxicity of plutonium, (b) The genetic effects of plutonium are negligible, (c) The carcinogenicity of plutonium is relatively inactive through all the routes of absorption except by inhalation because of the poor excretion of plutonium in that situation."

According to the *International Dictionary of Medicine and Biology* (1986) "Plutonium is chemically active and toxic." According to the *Encyclopedic Dictionary of Physics* (1962), whose contributors are overwhelmingly British academic and government scientists, including J. E. Gore, who contributed the article I quote, research into the physical properties of plutonium has been complicated by "difficulties in handling the metal, such as its high toxicity." So far as I can discover, no description of plutonium as other than chemically and radiologically toxic and as carcinogenic is reflected in reference literature.

The waste-disposal booklet says, "There is no proven instance of a human being suffering from plutonium intake." This is a favorite quibble. Leukemia, multiple myeloma, and lung cancer, though they are all associated with plutonium, cannot in any single case be proved to have been caused by it, supposedly, though they occur in high numbers in a plutonium-contaminated environment. Since each may have another etiology, so might all of them. Therefore, high cancer and leukemia rates cannot be said to be caused by exposure to plutonium, the causal link is not proven, and plutonium is exonerated. All this deserves its own chapter in any history of modern thought, simply because its consequences are epochal.

The booklet describes a regime of fastidious caution governing the operation of a "reference" plant, something between an actual, a hypothetical, and a projected system for fabricating nuclear fuel. "Windscale" is mentioned twice, never with reference to its accidents. In the midst of murky descriptions of contamination swabbed up with cotton and of the polishing of windows, under the heading "Special Problems," we are given a description of plutonium which concludes that there *are* no special problems. And yet the booklet describes how plutonium 241 will continuously decay into americium 241, with the release of beta and gamma radiation, so that "the specific external risk for plutonium operators is steadily increasing with time." The section concludes, "This may become especially embarrassing during waste handling operations." So even in claiming a good character for plutonium it puts aside the matter of its virulent decay product, for the management of which no guidance is offered, unless to be braced for embarrassment.

Why this little document should have been produced I have no idea, or how it could be read by anyone even moderately conversant with these issues and not inspire amazement and alarm. I think the official imprimatur may have dazzled skeptics. In any case, the radiological assumptions that governed British treatment of plutonium at the start of this decade are stated very openly and authoritatively here.

On the basis of the practices and assumptions described in these publications of Her Majesty's government, and taking into account the age and size of the plant, reckon what impact Sellafield is liable to have had. Add reports of radiation-associated illness, stern government inquiries with their implied disavowal of responsibility, and international protests, which in 1986 resulted in a vote in the European Parliament to close the plant down. The general purport of journalistic accounts, always to be considered suspect in their particulars, is confirmed.

If there is anything to the theory that a lump of plutonium the size of a grapefruit is toxic enough to destroy life on earth (I encountered this theory on the front page of The (London) Observer, in a chatty little story which included the information that the British government had poured that quarter ton of plutonium into the waters off the British coast) then the jig is up. I see no point in rushing to this somber conclusion, though it is very difficult to find any authority that will give plutonium a significantly better character.

West German animal tests are said to have demonstrated that a thousandth of a gram of plutonium killed dogs and rats in a matter of days. I have grown cynical enough to judge a piece of information by its effect, whatever its source. When a scientist declares that a speck of plutonium will kill a rat in a period of days, he is saying a spill of plutonium would create an unconcealable disaster. Then that part of a quarter ton which, over years, must have entered the environment and the food chain should have felled Britain by this time. The belief that overwhelming catastrophe would be the consequence of plutonium contamination implies anything less spectacular is proof that plutonium contamination has not occurred.

The official secrecy of Britain reflects the assumption that information damaging to the government should be contained where possible. Employees of the National Health Service must sign the Official Secrets Act, a fact which in this context sheds new light on the economic value of both the Act and the Health Service. There are no grounds for crediting public health data generated under such conditions, especially where it might inhibit a policy so enthusiastically pursued as Sellafield has been. Still, effects of radiation must have been limited enough, by the standards of the exposed population, to seem tolerable. Since Britain's industrial history has made occupational illness and injury commonplace, passivity relative to such problems is a settled feature of life. Effects will be more conspicuous over time, reflecting the cumulative and incremental enhancement of exposure which will come with further releases of wastes, and the decay of plutonium into americium, a more intensely radioactive element, and the continuing action of surf and tide in bringing the wastes ashore. Perhaps the limits of physical endurance will be reached before the limits of docility. That is usual in such cases.

To put the matter briefly, I am writing about the

radioactive contamination of a populous landscape—wastes from the plant can be measured on every coast of Britain—without special confidence in any description of plutonium, and without more than anecdotal evidence of the consequences of radiation exposure for public health. Grossly elevated rates of childhood leukemia and lymphoid malignancies in the area are conceded, though their significance is thought to be uncertain. In the Ravenglass Estuary near the plant, concentrations of plutonium are 27,000 times "background" levels, established by residues left from atomic testing which officials claim are high enough to create problematic contamination in the area, and which therefore must provide such calculations with a hefty multiplicand. Recently, officials have claimed to have no figures on levels of background radiation in Britain. Such discrepancies are commonplace. The Ravenglass Estuary near Sellafield once had a nesting population of 24,000 gulls and five other varieties of seabird. It is now virtually extinct. Eggs from the diminishing flock are radioactive, but no conclusion can be drawn from this fact.

As it happens, on the east coast of Ireland there have been numerous cases of Down's Syndrome and leukemia, and in England in the area of Sellafield, as fate would have it, houses have been found to be contaminated with plutonium. In response to the charges that the plutonium factory is to blame, both these phenomena were laid, by British officials, to the detritus of atmospheric testing, brought down by rain. Jonathan Schell, in *The Fate of the Earth*, remarks that there are such "hot spots," though he does not name any of them. Given the importance of British experts in the councils of the world nuclear enterprise, I wonder if their thinking is reflected in this hypothesis. I wonder if other "hot spots" are coincidentally centered around other nuclear facilities.

It is surely odd that ascribing health problems to the bomb tests would seem to anyone to exculpate Sellafield. If plutonium that falls from the clouds is harmful, then plutonium that comes in on the wind and the tide is no doubt harmful as well. After all, a woman in Cumbria who found that her house was contaminated when she sent her vacuum cleaner bag to the United States to be analyzed was obliged by law to sell the house at a very low price because it had a defect—the contamination. Apparently no law requires that the defect be corrected. One must conclude, however, that the British feel plutonium in one's house is something to be avoided or, failing that, regretted. Radiation is found in particularly high concentrations, one thousand times background levels, in household dust in the area, subsequent inquiry has shown.

I have suggested elsewhere that logic is not a ruling passion among the British. My problem in writing this apocalyptic tale in a style suited to the importance of its subject is in fact that there is a particular, somber, officious foolishness about it all, and a forthright miserliness which it was, until lately, my error to consider beneath the dignity of governments.

We are all accustomed to horror stories about nuclear plants. There are so many tales of near-catastrophe that they are almost soothing. Any dire information elicits a "Did you hear the one about—?" response, as if no story could be quite the worst, as if one's being impressed by any particular set of revelations is naïve, or a perverse sort of optimism, since it seems to assume that things work reasonably well anywhere.

British stories are of another order, however, because for them safety as we understand it was never a primary objective. An anecdote is related from time to time about how Clement Atlee, Prime Minister at the end of the Second World War, ordered that Sellafield (then Windscale) should have absolute priority in obtaining building materials. But the instruction was marked "Top Secret," so no one was allowed to read it, and therefore the builders had to compete for materials without any preference shown to them. For the British, the idea that government conduct is based on fluke is exculpatory. The government has no positive obligation to see that a policy is carried out competently or humanely, or to correct its past failures. It strains belief that a country which had just come through a war, and which had lived since 1911 with the Official Secrets Act, under which the whole of "public" business has the status of military secret (the better to confound the Hun), could not handle a straightforward matter of military procurement. Since Windscale/Sellafield was a munitions manufacturing site before it became a plutonium factory, mere inertia should certainly have assured it some preferential treatment. The anecdote does at least acknowledge that the plant was and is shoddily built, and a great deal of evidence is available to prove that this is indeed the case.

The report of a recent inquiry into cancer deaths near the plant remarks gently, "While there is ample evidence of a real and sophisticated concern with the safe operation of the plant, it has to be said that some of the plant was installed many years ago."[*] James Wilkinson, science correspondent for BBC-TV, finds French waste

---

[*] "Cautious view of Sellafield 'cancer link,' " *The Guardian*, July 24, 1984, p. 4.

disposal methods "tidy," while British arrangements are astonishingly "tatty."*

Oddness and awkwardness surround every transaction that involves information in Britain, whether the taking of it in or the giving of it out, as the Atlee anecdote suggests. There is clearly a deep cultural ambivalence about information. It is a power vested in the innermost counsels of the state. And then the state can claim to have failed to generate it, as in the matter of health data and radiation levels, or it can reject information it finds not to its liking, as in the matter of recent reports of structural flaws at these old nuclear sites. The plants were designed to operate for twenty years. In 1976, when the first were due to close, the Labour government in the person of its telegenic Minister of Environment, the awfully ferocious radical and pretty persistent gadfly of entrenched interests, Anthony Wedgwood "Red Tony" Benn, after a lot of study, decided to keep the plants running. This is perhaps an instance of Benn's ability to set his populist enthusiasms to one side.

Despite the implications of design limits and the building of these postwar plants out of materials about which no one seems prepared to say anything kind, except that they were inexpensive, official statements of the risks these plants pose boggle the mind by invoking the far reaches of the infinitesimal.

"Odds" always imply, amazingly, that there is some sort of safe landing on the other side—a plant might start up, waltz through its productive life, skirting every danger, and then be done and gone. The time that has passed from the fall of Troy to the present day

---

* "How safe is Sellafield? Claims about nuclear leaks should be handled with care," *The Listener,* February 27, 1986, pp. 2–4.

will not bring humankind halfway to the end of the problems any one of these plants creates in the course of quiet, impeccable functioning. Consider for example that calculations of the warming of the earth and the rising of the sea have Sellafield and a great many other sites under water in the next century. Reactors are called "breeders" and nuclear materials are described as "fertile," and in fact the functioning of a reactor even in the best circumstances is the gestation of ferocious elements this suave little planet was never meant to contain. Tapping electricity from such a phenomenon is like setting the house on fire to toast marshmallows.

Since Calder Hall has been singled out for faulty design, there is special cause for worry. This old reactor, opened by the Queen in 1956 at Windscale and said by the British to have been the first reactor in the world to generate electricity on a commercial scale, does, after all, share a site with the largest repository of nuclear waste in the world, and nuclear wastes differ from egg shells and potato peelings in that in storage they must be constantly and solicitously tended. They are simply reactor cores that have become too fissile, too radioactive, to be used to generate electricity. To describe them as "spent" is entirely misleading. Like reactor cores they generate enormous amounts of heat and require continuous cooling. This failing, they burn voraciously, through concrete. The Russians were able to cover over their damaged core and to contain the release of radiation by burying the reactor in lead and boron, for the time being at least, or so we are told. But the effort required was gigantic. The problems involved in keeping radioactive wastes safely stored are about the same as in keeping a reactor stable. So a major accident in a waste storage site, where tons of reprocessed plutonium are stored also, would set in train a series of consequences that could put all previous misfortune very far in the shade. As the Russian events made clear, a nuclear accident is difficult to contain because it creates circumstances in which it is hard to keep other reactors at the same site under control. At Sellafield, the problem would be many times compounded by the nature of the establishment.

In 1958 the Russians had an explosion in a nuclear waste dump at Kyshtym in the Ural Mountains, a not especially famous disaster that took towns from the map, depopulated hundreds of square miles of countryside, possibly killing hundreds of people, according to the Russian physicist Zhores Medvedev, and filled Russian scientific journals with distinguished contributions to research into the impact of cesium 137 and strontium 90 on biological systems. Russia is very large, and has growing reason to be glad for all that space. Britain cannot sidestep the full consequences of its errors, nor can Europe pull back its skirts from the mess that will be made if any such thing should happen in England.

If Americans have heard about the Sellafield nuclear waste dump and plutonium factory, they have heard the name Windscale, which appears from time to time with little or no elaboration in lists of nuclear accidents. The Windscale fire of 1957, which for our purposes is the history of the public-relations strategies surrounding the event, bears an uncanny, not to say unnerving, similarity to the recent accident in the Ukraine. Windscale was the most serious accident in a nuclear reactor before Chernobyl. It occurred in a graphite-moderated reactor with the sole function of producing plutonium for British bombs. Its causes were promptly located in "human failure" rather than in any problems with the design of the reactor, which was said before the fire to have provided the model for British power-generating reactors, and after the fire to resemble them hardly at all. An inquiry found that staff had undertaken an experiment at a time when the reactor was going through a routine maintenance procedure, just as their Russian counterparts did thirty years later. The nature of the British experiment has never been revealed because it was defense-related. On these same grounds the real composition of the radioactive fallout from the accident has only gradually begun to be acknowledged. According to new accounts, it was very much more virulent than originally claimed.

Local reports of the event described an explosion, a fire, and a massive release of radioactivity. Official reports described the core temperature rising to red heat, and a release of radioactive gases mainly trapped in filters above the pile. Subsequent revisions of the official account concede that there was indeed a fire, and a cloud of radioactivity, which contaminated England, Ireland, and Europe. No one was evacuated. A reassuring press account describes children in the nearby village of Seascale playing in the streets.

Comparison in this regard is to the advantage of the Russians, who only delayed evacuation, and who only temporized for a few days about the severity of their problem. Satellites and monitoring devices may enforce candor, at least at the discretion of the governments or firms who operate them. In 1957 an extraordinary degree of concealment was possible. One "casualty" was acknowledged, in quotation marks because he was only contaminated, and was to be seen the next day wearing rubber gloves while playing dominoes in a pub. He was said to have been among those workers who trained fire hoses directly on the burning reactor core when other attempts to cool it failed.

Men from other facilities were brought in to work in forty-minute shifts. Some were said to have collapsed, but officials denied this robustly. They declared that no one had suffered any harm. The fire in the reactor core was out of control for two days, even according to early reports. The pile had been functioning for seven years, so its inventory of radioactive materials would have been much more virulent than that in the plant at Chernobyl. There was no containment structure. Dousing such a fire with water would inevitably produce, at best, radioactive steam in enormous quantities. Water was poured into the core for twenty-four hours.

The physicist in charge of the routine maintenance operation was found to have had no manual of instructions for carrying it out. This was seen as a collective responsibility and no disciplinary action was taken. So, as at Chernobyl, an extraordinary concatenation of misjudgments produced an accident which had no implications for the nuclear industry as a whole. As at Chernobyl, amazing good fortune prevented the consequences of disaster from being as severe as might have been expected, at least according to the newspapers. Prime Minister Macmillan assured the Parliament weeks after the Windscale accident that there was no evidence of harm to any person, animal, or property. This is remarkable, considering that milk produced in a 200-square-mile area around the plant had been confiscated and poured into, of course, the sea, during those same weeks. Now the British attribute about 260 cases of thyroid cancer to the accident. Perhaps in 2017 the

Russians will also revise their original estimates of the seriousness of Chernobyl.

In this early event features highly characteristic of British handling of nuclear issues are already fully apparent, not least typical being the singling out of thyroid cancer as the one result of a reactor core fire. This follows logically on the pouring out of milk as the one measure settled upon to protect the population at the time of the fire, in its turn a consequence of emphasis on the release of radioactive iodine. All sorts of things would have come from a plutonium-producing pile in which graphite and uranium burned for days, and there would have been an array of aftereffects. But thyroid cancer is said to have a high cure rate, so that only a few deaths need be attributed to the fire if this kind of cancer is treated as its only consequence. Cancer of the breast or lung, also radiation-associated, would imply many more deaths. It is perhaps not irrelevant to note here again that Britain leads the world in lung cancer deaths.

Recent concerns about the consequences of exposure to radiation focus, just as arbitrarily, on clusters of childhood leukemia, which seem, for the purposes of those who document them, to mean rates of mortality which exceed national averages by about 1,000 percent. In any but the grimmest circumstances, such excesses should be relatively rare. In fact, they have been found near many British nuclear sites. This phenomenon is usually associated, speculatively and again arbitrarily, with environmental exposure to plutonium, though there are many other radiation sources in the environment. The association of foetal X-rays with childhood leukemia demonstrated by the British physician Dr. Alice Stewart indicates that even a brief, discreet exposure to radiation is sufficient to pre-dispose a child to this illness. Narrowing of the terms in which a problem is to be understood remains a flourishing art.

Fully present also in the Windscale affair is the tendency to treat every problem as a public-relations problem first of all. Asked in Parliament whether he would publish the government report on Windscale or support a proposal for further inquiry, Prime Minister Macmillan agreed to consider these steps, "but he was also interested in maintaining the tremendous and unique reputation of our scientists in this field through-out the world," according to the *Times* report of parlia-mentary discussion of the accident.[*]

Public safety, where it is in conflict with prestige and export prospects, has no standing. This exchange was reported about three weeks after the accident, when even short-lived contaminants would still have been present in significant amounts. Yet the public was told from the first it was safe to eat local vegetables and to let cattle graze in the open air. These assurances came in the face of protests by farmers and by construction workers employed at the site.

Twenty years later, at the time of the Windscale inquiry of 1977, which yielded the decision to expand the plant's role as waste dump and plutonium factory, half the mortality data on workers was found to be missing from the files.[†] Those that existed were thought sufficient to provide an estimate of the plant's safety. There is no reason to doubt that the prestige of British nuclear development did indeed emerge unscathed

---

[*] "Early statement on Windscale," October 30, 1957, p. 5.

[†] "Windscale to check deaths records," Michael Morris, *The Guardian*, July 14, 1977, p. 5.

from the events at Windscale, and every reason to look for its influence in other quarters, especially on our own industry, which has served as a catch basin for the brain drain.

The American trade publication *Nuclear News* reports respectfully on the affairs of the British industry, alluding to radioactive discharges and emissions with a serenity I can only find alarming, since its readership is usually represented as earnestly concerned with preventing measurable "radiation doses" to the public from nuclear sites. British assurances of the harmlessness of such exposure are reported at length, without any hint of skepticism, and without any of the detail or specificity a lay person might hope the specialist community would demand.

After Chernobyl, a British report, highly critical of the Russians, was described in an article which quoted Walter Marshall, head of the Central Electricity Generating Board, as saying no such accident as Chernobyl could occur in Britain because "the overriding importance of ensuring safety is so deeply engrained in the culture of the nuclear industry that this will not happen in the U.K."[*]

There are said to have been three hundred accidents at Windscale/Sellafield since the core fire. Such figures are meaningless. Accidents are creatures of definition. An industry that ignores every standard of caution is almost proof against accident. There have been events that required buildings to be closed and abandoned, there have been fires. There was once a flood. None of these approach the normal functioning of the plant as sources of contamination. But a history of accident gives

---

[*] "UKAEA Report Indicts Soviet RBMK Design," *Nuclear News*, August 1987, p. 68.

the place a kind of respectability, implying standards and scruples.

Britons figure prominently in world organizations which generate standards the British violate with a special flagrancy. How do these gentlemen resolve the contradictions of their national and international roles? John Dunster, health physicist at Sellafield in its formative stages, early defender of its plutonium dumping, has since become one of the two longest-serving members of the International Commission on Radiological Protection, which establishes exposure standards for the Western world.

Such contradictions abound. It seems there is a branch in Cumbria of the Medical Campaign Against Nuclear Weapons, an organization of doctors appalled by the darker potentialities of the age. They meet in premises available to them, according to an article in the *New Statesman,* on the condition that they say nothing against Sellafield.*

An editorial in the British magazine *New Scientist,* accusing the American press of fueling anti-Soviet hysteria by demanding information about the accident at Chernobyl, remarked, "If the Soviet authorities want their people to die in ignorance, then it is up to them."†
The issue of secrecy opens on the very largest questions of legitimacy and social order. There is a consensus in British life that keeping up appearances is a thing to be done at any cost. There can be no doubt that the appearance of reasonableness, morality, and good order is deeply important to them, of great value in its own right. Pervasive as secrecy-enforcing laws and

---

* "Fitting in with the locals," Ruth Balogh, September 14, 1984, p. 8.
† "Nuclear propaganda," May 8, 1986, p. 16.

conventions are, no one ever seems to refuse to talk about anything. Secrets are kept by saying something other than the truth. According to an article in *The Guardian*, a report on the Windscale fire published in 1983 by the National Radiological Protection Board described the original report by John Dunster, its head until 1987, as "spurious."* I am describing a very remarkable culture.

Putting aside such obvious mystifications as missing data, other issues seem to lurk behind the Windscale accident which would make its significance, great as it was, only relative. A Labour MP, asking for explicit assurances that the drinking water for the city of Manchester, which comes from lakes in Cumbria, was safe, said, "Before the accident happened some very strange things had been taking place in the Windscale area, though no public information has been given."† An article in *The* (London) *Times* addressed the issue of nuclear waste disposal, which seems to have been implicated in the public mind in the events at Windscale. The article explains patiently that the two were not related, that the siting of the reprocessing plant near the reactors was merely convenient and had no bearing on events at the reactors. It was just at this time that complaints arose at the United Nations about plutonium dumping from Windscale. British scientists experimented with incinerating plutonium in Australia. I suspect that some such toying with disposal methods

---

* "New Windscale report hints at 33 deaths," *The Guardian*, September 28, 1983, p. 3.
† "Inquiry ordered at Windscale," *The Times*, October 16, 1957, p. 10.

may have gone on before the accident, whether or not such experiments were connected with it, and that the contamination from the core fire may simply have frosted the cake.

So early on, the assumptions upon which Windscale/ Sellafield would operate were already firmly established. In the article about waste disposal published after the accident at Windscale, *The Times* describes the division of radioactive materials into three groups: those that are sufficiently short-lived to be managed by a limited period of storage; strontium 90 and cesium 137, which are too long-lived to be treated in this way and for which commercial uses should therefore be found; and a few materials "so long-lived that the amount of radiation they contribute is not significant."* By this must be intended plutonium and uranium, both low-activity and long-lived. The article explains that "as a matter of common sense rather than science, it has been fairly generally accepted that so long as the total additional radiation is small compared with natural radiation the amount of harm done, if any, is unlikely to be detectable." Especially if health data turn up missing.

The British approach to the disposal problem is sufficient to deal with the entire world output of waste, according to *The* (London) *Times* in 1957. Fully twenty years before the decision was made public, the commercial or scientific (or common sense) rationale for Sellafield was already fully formed. That is to say that, in twenty years, the thinking of those responsible for the operation of the plant did not develop.

In 1977 Dr. Geoffrey Schofield, chief medical officer at Sellafield, testifying for the plant's expansion, said

---

* "Disposing of radioactive materials," October 19, 1956, p. 15.

plutonium had become "a great bogeyman."* It would be tempting to say that the cleverest class of the cleverest nation wagered on the innocuousness of long-lived fission products in broadcasting them over the landscape, if it were not the case that they have always permitted flagrantly damaging materials, specifically strontium 90 and cesium 137, to pour into the environment as well. Interestingly, an essay titled "The Guarantee of Safety: Protection Built In," by the Group Medical Officer of the Atomic Energy Authority, included in the supplement published by *The* (London) *Times* a year before the accident, when Calder Hall was opened, said, "It is important to note that the British nuclear power programme has been so planned that these chemical processes [that is, reprocessing] will be carried out at one or two sites remote from the power stations; by this arrangement, the complicated problems of protection against the risks of radioactive contamination will be isolated in the chemical processing plants and will be excluded, therefore, from the nuclear power stations."† Reprocessing at Windscale/Sellafield has never been remote or isolated. As this passage implies, it has never been designed to obviate contamination by any other means.

The issue of the super-addition of man-made to natural background radiation still arises, still in the same terms. If an increase in radiation in the environment is small as a percentage of background, no harm has been done. Now, if this were believed in good faith, surely there would be publicly available figures for background radiation to serve as a basis for measuring any increase. Such figures are said not to exist. Complicating factors

---

* "Plutonium threat 'a bogeyman,'" *The Guardian*, July 15, 1977, p. 3.
† A. S. McLean, *Calder Hall Supplement*, October 17, 1956, p. x.

are adduced; for example, that levels of natural radiation vary from place to place. But then measurement devices can be moved from place to place. Weapons testing has enhanced radiation levels in the environment. But if radiation levels were established, the impact of other sources could be measured nevertheless. If public-health decisions are made on the basis of supposed safe rates of increase of radiation exposure, it surely behooves those who make such decisions to supply themselves with a basis for their calculations. Lately the meaning of the word "background" is shifting, so that it refers simply to whatever is there, with the implication that existing levels are safe. What constitutes the area to be described as "background"? If the only significant source of radiation for a mile in every direction is the fish in my soup, what comfort can I find in the fact that, averaged out over some unspecified area, the radiation "dose" is insignificant?

These calculations are rather insanely abstract. Contrary to vivid experience, Sellafield apologists seem to imagine that the wastes put into the environment are in fact "dispersed." So they speak often of "undetectable" harm, the kind that is owed to the plant but cannot be attributed to it. For them the word means something very like "nonexistent." Sellafield officials are reported to have asked the area county council—in which Sellafield is well represented—for permission to raise emissions.* The new levels would produce, by the reckoning of the council's radiation expert, one extra cancer a year. It was agreed that this undetectable death would probably occur in the population near the plant. Permission was granted. Of course that was when the

* "The peril that lurks by the sea," James Cutler, *The Guardian*, July 24, 1984, p. 15.

death from cancer of one child in sixty in Seascale, the nearest village, was still undetected. Radiation-induced cancers among the Sellafield work force are undetected, because the plant management has never accepted that any cancer has resulted from working there, though it admits contamination is commonplace.

The use of Cumbrian lakes as reservoirs for cities sheds an interesting light on the practice of expressing cancer rates as multiples of regional or national rates. The runoff from the mountains would surely concentrate contaminants in the drinking water of large populations. The nearness of Sellafield to Manchester and Liverpool is seldom alluded to, but there are hundreds of thousands of people living within the range of its effects. So far as I can discover, illness in such populations only serves to make health anomalies nearer the plant appear less exceptional, harder to "detect."

Information, for want of a better word, is always suspect, and it is continuously undercut. On page 1 of *The* (London) *Times* of May 21, 1985, there appeared an article titled " 'Plutonium food' sought for children," an article which fairly epitomizes the complexities of following this issue in the British press. It was said to be based on leaked minutes of a meeting of representatives of government agencies whose duties and expertise were relevant to the situation at Sellafield, including the Department of Health and Social Services. The stolen minutes were supplied to a parliamentary committee by Greenpeace. At this meeting it was allegedly proposed that "volunteer" children should be given plutonium-laced food to see how it affected their bodies. The proposal was roundly denounced by the committee. One of those present is reported to have said, " 'How many parents would volunteer their children? Are we living in the real world?' "

The DHSS should have whatever data there might be about the human impact of radioactive contamination in Britain. While one is struck by the low level of moral refinement that would be reflected in the notion of feeding a toxin to children to observe whether and how they are poisoned, stranger by far is the apparently profound naïveté reflected in such a suggestion. To act as though no information on the subject exists, and that the way to develop information would be to perform an experiment on human children, is truly remarkable, though this latter may reflect a sensitivity to the views of animal-rights groups, which are mighty in Britain. The *Enclyclopaedia Britannica* has a thing or two to say about the toxicity of plutonium, and there is a scientific literature of some quality and interest which should be known, or at least known to exist, by individuals who have made plutonium their stock-in-trade.

In this case it was subsequently reported in *The Observer* that Greenpeace had been "condemned by the House of Commons Environment Committee for falsely alleging that it had been suggested that Cumbrian children be given food contaminated by radioactivity."* In the same article, for good measure, it was also reported that a former Greenpeace spokesman had signed on as advisor to a new scheme for depositing wastes in the sea floor. In light of the peculiarities of British press and secrecy laws, I think it is always reasonable to wonder about such things as "leaked

---

* "Atomic protest could backfire," April 26, 1987, p. 4.

minutes," because it seems extremely likely that any leaked information is actually planted. For why should the publication of such information not bring down penalties, when, under the same government, the offices of the BBC and the *New Statesman* have been raided and searched by the police, *The Observer* has been forbidden to allude to information contained in certain of its own articles (having to do with a book by a former MI5 agent), and so on? The authenticity of the reported suggestion is said in the *Times* article to have been confirmed by the managers of Sellafield, British Nuclear Fuels, who explained that "the idea of feeding children contaminated food was not a serious suggestion; it was a throw-away remark." The chairman of the committee and a Conservative MP are both named in the article as harshly critical in their reaction. It was the chairman who put the question about living in the real world. Perhaps the government wished to distance itself from BNF, in preparation for new management and implied reform. It may be that this select committee wanted to get its ignorance on record—where ignorance is exculpatory 'tis folly to be wise. The subsequent renunciation of the minutes does no more than to cast doubt on their authenticity, nuancing the effect, so that the question of competence can remain unresolved.

But the oddest thing of all is that, as the *Times* article implicitly acknowledges, the feeding of plutonium to children living in the area of Sellafield is entirely redundant. The children there have already been fed plutonium. Another suggestion reportedly offered at the meeting was that "placentas and still-born children should be analysed for concentrations of radioactivity." Reports in other sources indicate that this is in fact being done. Another article in *The Times* reports confirmation by officials that aborted and stillborn babies, afterbirths, and children who die in accidents are tested for plutonium concentrations.* Children have been experimented upon *ab ovo* and, together with shellfish-eating fishermen who have been used for "monitoring," they constitute two ends of a continuum whose intervening stages might be inferred. And of course there are the children dead of leukemia, whose peers and siblings surely bear watching, and all the presumptive plutonium eaters of Cornwall and northern Wales and northern and western Scotland and Northern Ireland.

It surely would be strange to respond to having fed plutonium to children with the suggestion, even as a "throw-away remark," that plutonium should be fed to children. To propose that the thing should be done is to deny that it has been done already. To renounce the report that such a proposal was ever made intensifies this denial. Then all the outrage directed against what is as if merely suggested or alleged puts those who act out disapproval in a proper moral position. The parliamentary committee meeting is like a ritual in which reality is magically altered, evil is resisted, and sanity affirmed, an effect reiterated in the subsequent, quite plausible denial that this remarkable proposal was ever made. But as I write and as you read, plutonium is flowing into the human environment, courtesy of this same government.

Perhaps I should use my own reaction to interpret this artifact, this putative stolen glimpse into epochal deliberations, reported in a reputable newspaper favorably disposed to government and industry. At first I thought, These people are very foolish, very ignorant, and, I imagined, dabblers, sheltered incompetents. I

---

* " 'No secret' over tests on Sellafield babies," July 29, 1985, p. 3.

forgot that, when the huge state and private companies which have marketed nuclear construction and equipment, as well as radioactive materials, and have undertaken the transporting of wastes—and the supply of expertise—are taken into account, the British nuclear industry is old, vast, highly elaborated, and profoundly influential. And I forgot that these people have pulled off the public-relations coup of all time, inverting every rule, pouring out "routinely" all the toxins we are always assured no foreseeable accident could release from any reactor, doing so without qualm or hesitation, without any loss of face or of moral confidence. Only think how many people outside Britain have known this for decades and never said a word.

To compete with great success in a sophisticated industry requires and implies sophistication. This success is based not on technical competence so much as on an amazing gift for concealing the obvious. And the trick is working even now. Why should the food given hypothetical "volunteer" children be laced with plutonium? Why should the autopsy tissue of actual children, who were in no sense volunteers, be tested for plutonium? Why not radioactive iodine, which is vented from the plant's stacks? Why not cesium 137, which flows into the sea, contaminating meat and milk and fish? Better still, why not all three, with a lashing of chemical solvents that break down reactor cores and a soupçon of the uranium dumped daily by the chemical plant up the coast? Plutonium is only an aspect of the problem, and not the most acute by any means. If the emphasis on it is intentional, it is shrewd, because it tends to simplify and therefore minimize the problem.

Bother about the stolen minutes makes the alleged suggestion sound more untoward than it really is, measured against other reported policies. An article published in *The Guardian* titled "Radiation tests ordered on Sellafield food" describes a plan by the government "to take samples of food to test whether it has been contaminated" by a recent series of leaks from the plant.* The tests were to be coordinated by the National Radiological Protection Board, the incredibly feckless agency responsible for monitoring public exposure to radiation. The point of the study will be "to try to establish *how much* radioactive waste has entered the food chain" (italics mine). The design of the study is very interesting. "A sample of the local population will be asked to buy twice as much of the food they normally purchase over a seven-day period." One half of the food "will then be analysed in government laboratories." Clearly the object is to establish levels of intake. Otherwise food would simply be tested at random. In other words, the thing to be established is how much "radioactive waste" is eaten by individuals—what tastes or habits or budgetary considerations might result in high levels of ingestion. Ethically, this is more than a little similar to the disclaimed suggestion that, in *The Observer*'s words, "Cumbrian children be given food contaminated by radioactivity." While children are neither explicitly included in the later study, nor excluded from it, they are among the population whose food is presumed to be contaminated. And since they are not prevented from eating it, or provided other food, whether their exposure is noted by scientists or not is a question of very little interest.

And still there is the matter of singling out food as the

* "Radiation tests ordered on Sellafield food," Richard Norton-Taylor, February 21, 1986, p. 30.

"pathway" of radiation exposure. The spills into the sea and air could impinge in all sorts of ways. More to the point, as usual they are treated as novel and alarming, though the whole landscape has been exposed to accumulative contamination for decades.

The 1976 *Nuclear Energy in Britain,* the British government publication mentioned earlier, describes as follows the work of the Biology Department of the National Radiological Protection Board: "In order to derive appropriate standards for air concentration and maximum permissible body content, the distribution, retention and excretion characteristics of these materials [plutonium, americium, and curium] following their intake by experimental animals is studied." Also studied are "the mechanisms of production of chromosome aberrations by ionising radiations." Interestingly, these passages do not appear in the 1981 edition of the same pamphlet. They certainly indicate that the government has sponsored research by the NRPB of a kind to clarify the issue of the health effects of plutonium contamination. Yet the government claims to have no such information, as for example in the case of the Black Report on the leukemia deaths of Cumbrian children in 1984.

According to the 1976 edition of *Nuclear Energy in Britain,* "the most important route of entry of radioactive substances into the body is by inhalation and deposition in the lungs." Yet inquiries always imply that ingesting these materials would be the most important source of exposure to them, even in an environment where household dust is radioactive. Again, focusing on one "route of entry," like focusing on plutonium, makes the contamination seem much less complex and pervasive than it obviously is.

*   *   *

Reading over the news of decades, one grows accustomed to certain faces, which appear from time to time in the murky chiaroscuro of photographs like fish nosing up against the side of a tank, a little startled to be reminded again that there is a world to which they are of interest. The fine, mild face of John Dunster appears from time to time. I saw it recently on my own television screen. He and other British worthies had been enlisted to explain the nature and consequences of the events at Chernobyl to the American public. They were gravely reassuring.

The most recent news I have of John Dunster is an article from the British magazine *New Scientist* by Michael Kenward, describing yet again the work of the National Radiological Protection Board.* The *New Scientist* is always of interest, being widely read in America, and being a trove of cryptogrammatic journalism. So an article on the NRPB is of interest for a number of reasons.

The NRPB, we are told, is pinched by lack of money. British journalism often adopts a cozy tone in discussing parsimony, especially where it is seen to abet incompetence in the government. A recent role filled by the NRPB, after Chernobyl, was "the matter of advising the government, few of whose ministers knew an isotope from their elbow." This same government has been running a nuclear rag-and-bone shop for decades. Considering the very great decisions, great in terms of their consequences, that Britain has made for the whole

---

* "The nuclear watchdog strains at the leash," May 22, 1986, pp. 58–59.

world, in massively polluting the sea, in producing plutonium in quantity, and in treating the stuff of doom as an article of commerce, it has taken upon itself a positive obligation to be capable of even finer discriminations. The NRPB is, according to the article, a small, besieged organization, which lost fifty staff members to budget cuts between 1980 and 1982, and which still struggles with a shrinking budget. It was obliged at the time of Chernobyl to inform its government of the consequences of radioactive contamination. No one in Whitehall, no advisor for the Ministry of the Environment or the Ministry of Agriculture, Fisheries and Food, entities responsible for approving waste dumping and emissions—and they do approve, heartily—none of these people was competent to describe the nature of radioactive contamination.

This may seem odd, among the cleverest class of the cleverest nation in the world, yet I am very inclined to believe it. Rather than being seen as reducing the government's authority, ignorance actually seems to bolster it, amateurism being a term of praise among these people. There is really no reason to imagine the Minister of the Environment swotting up on the mutagenic properties of alpha-emitters when the possession of such information could only bring regret, given that the British environment is already salted with them and that the government is committed to a policy of always greater waste accumulation in that densely populated island, and that neither law nor custom demands competence from him, even of the most general kind, any more than they expose him to the painful obligation to speak the truth in these matters of pressing national and international interest.

The article has a tone of having taken a deep breath

and made a fresh start, which invites readers to forget
what they may have read yesterday. After all, at the time
of this article, British nuclear issues had been filling the
press, including the *New Scientist,* since the fall of 1983,
about thirty-one months, and had arisen in pressing
forms at intervals since the late fifties. Yet it is as if
Britain were an enchanted island, its government
aroused from Keatsian indolence only by an alien east
wind. In fact, the government at first insisted Chernobyl
had not had a significant effect on Britain. And, to
clinch the argument, it did not test for contamination.
Or it was testing, but a rain began to fall and the NRPB
had to move its instruments indoors to keep them from
getting wet—another *New Scientist* version of the story.[*]
Or it tested only in England, one cauliflower and one
cabbage in the northwest, one sample of spring greens
in the center, elsewhere "two of parsley, four of spinach,
four of cauliflower, one of rhubarb, fourteen of cabbage, five of broccoli, four of eggs, and a single trout,"[†]
like supper in a Peter Rabbit story. In an article published before the extent of contamination came to light,
the *New Scientist* reported matter-of-factly that "people
would have received, on average, 0.008 millisieverts
extra in one week from radioactivity in the air . . . 0.002
millisieverts from drinking tap water for one week . . ."
etc. These calculations to the third decimal place came
from the Ministry of the Environment, the parsley
testers, and were presented under the title "A lot of fuss
about a few millisieverts."[‡] Information seems to have a
very short half-life. For weeks the British bought and

[*] "Ackworth takes charge," Grenville Needham, June 12, 1986, p. 58.
[†] "Introducing the national cabbage monitoring network," Ian Mason, *New
Scientist,* May 22, 1986, p. 23.
[‡] Sharon Kingman, May 15, 1986, p. 26.

consumed highly radioactive food, specifically lamb contaminated because the pastures were hot. Then hundreds of thousands of sheep and lambs were removed from the market—temporarily, of course. Need I say, the areas along the Irish Sea and in Scotland—coincidentally, those most affected by radioactivity from Sellafield and Dounreay—proved to be especially badly contaminated?

Interpreted as it is in Kenward's article, the delay in government response is attributable to the outré and uncongenial character of the circumstances these ministers are suddenly obliged to deal with. There is no allusion to the oddness of the fact that the government does not routinely monitor radiation levels around its two most important and controversial installations, though both have histories of accident and both have been associated with elevated cancer rates, and both have been the subject of international protests.

It will be remembered that Chernobyl came to world attention first because it set off an alarm in a nuclear plant in far-off Sweden. If radiation came down with special vengeance in the regions of these two British nuclear installations—which are represented as technical miracles except when they are being defended as old and shabby—surely their alarms should have responded at some point and the fact should have been reported to the British government, since every government in the Northern Hemisphere was watching the effects of Chernobyl in those days.

One may imagine alarms that never sound, or alarms that sound so often they are not attended to. However it may have been, Britain declared itself unaffected by the cloud from Chernobyl, during the time, perhaps, that the NRPB was describing to ministers the prevalence and health implications of the elbow. The article says that in the days after Chernobyl this "radiation watchdog" worked overtime "to supervise the monitoring of air and rain from the east." They must have forgotten to switch something on.

The article that has set me to pondering is a comprehensive description of the work of the NRPB, which the organization itself was outlining in a "corporate plan" just at the time of Chernobyl. (I must ask my reader not to be put off by acronyms. I am persuaded, more or less, that they exist to repel inquiry, to suggest expertise, and also to suggest drab toil taken on for the rest of us by those whose blood is thin enough to make tedium congenial, and to starve out the full-blooded vice and full-blown madness that might rage among others entrusted with such great responsibilities. Such officious nonsense is a DO NOT DISTURB sign hung on doors behind which things are transacted which should disturb us all.)

John Dunster himself is quoted describing the organization he directed thus: " 'The board aims to establish an overall policy in radiation protection that provides a proper standard of safety without unduly hampering the beneficial practices giving rise to exposures.' " In other words, it exists to adjust exposure standards to " 'beneficial practices.' "* The director of the NRPB, longest-serving member of its international counterpart, believes that practices giving rise to radiation exposure can still be beneficial. One craves elaboration. Yet there would be little value in it. As the article notes parenthetically, "(It isn't the NRPB's job to decide on the government's policy, so, as Dunster admits, ministers are quite free to ignore the board's advice.)" So the policy is

---

* "The nuclear watchdog strains at the leash," *New Scientist*, May 22, 1986, p. 59.

merely "advice," which "ministers"—who, as we have just been told, tend not to know isotopes from elbows—can ignore. This is most wonderful.

Writers on government sometimes remark that Britain has had trouble generating a concept of the state. The problem of the state of Britain is that while on the one hand nothing seems really independent or distinct from government, on the other hand, when responsibility is to be located, the government recedes like a dream and is nowhere to be found.

We have seen in Michael Kenward's article how the curtain rose on ministers all unsuited to deal with nuclear issues at the time of Chernobyl. In 1970, we are told, the government set up the NRPB "when it realized that all of the organizations in Britain with any knowledge of radiation and its effects on people were part of the nuclear establishment and were, therefore, compromised in the eyes of the public." This nuclear establishment was and is a creature of the state, funded, shielded, and patronized by the government, and flourishing in the balmy atmosphere of Crown Immunity, where no acts of Parliament apply, and under the protection of laws affecting national defense and commercial confidentiality as well as the Official Secrets Act, and under the supervision of ministers who have not made themselves competent in the area of their responsibility, and who are very much inclined to waive such standards as the government pretends to. If the United Kingdom Atomic Energy Authority, which carried out the policies of the government, became discredited, why was the government itself not discredited? Why does it have no obligation to generate and enforce standards of practice for which it will answer to the public? And what *is* a government, in any case, that can shed a compromised institution like a dirty shirt? When the UKAEA was abandoned in favor of the NRPB, the new entity was staffed from the ranks of its predecessor, with John Dunster as its head. British Nuclear Fuels, which now operates Sellafield, was created in the same way for the same reasons. The plutonium factory itself had its name changed from Windscale to Sellafield to rid it of an evil reputation. Recently that company's management was replaced, because the government lost confidence in it. The new management immediately retained the services of a public-relations firm. It now shares advertising space with Greenpeace.

This changing of letterheads and images neither promises nor effects any change in government policy or industry practice. "Practice" at Sellafield merges quite indistinguishably with accidents and spills and faulty record keeping, in any case. So, while there are no meaningful standards, the greater problem is the un-suitability of the facility to comply with standards should they exist. Again, this hapless plant is and always has been the exclusive property of the British government.

When Henry VI brought England to ruin in the fifteenth century, there was some debate as to whether the King was a saint or an imbecile. In either case, since he could not be expected to understand the evil that went on around him, he was not to be faulted for it. The British government is, as an idea, some such specter of irreducible innocence and sanctity, liable to being badly served in the very degree that it *is* innocent. This could very well be nothing more than the survival of some especially successful medieval public-relations cam-paign. Why old myths should be assumed to be more respectable in their origins than new ones I do not know. The point is simply that the phantom sovereign, always to be revered and never held to account, is a major part of the phenomenon of political sovereignty in modern Britain.

Yet clearly the government—"ministers," in the terms of Michael Kenward's *New Scientist* article—is always decisively involved in what happens. It reserves to itself truly extraordinary powers. For example, the NRPB corporate plan will give the government an idea of the board's plans and activities "in words of as few syllables as possible," that is, appropriate to the understanding of imbeciles. Yet the reason for preparing a plan the ministers might understand is to allow them to override it: "If the government wants to point its watchdog at new scents, then at least it has something to go on, some false trails to abandon." So this "independent" watchdog agency is to allow its agenda to be set by the government, which is also the nuclear industrialist and trash collector.

The NRPB is reported to be "embarrassed" financially and hinting obliquely for a little increase in staff. This is poignant. It is always touching to see how British institutions struggle against the hard demands of government thrift. It is also an instance of the enormous power of the government to expedite its policies by omission. By failing to fund its monitoring agency at an adequate level, it is preparing a defense for itself and the agency as well, in the event that things are done now which will someday have to be disowned. The government is rendering itself less competent, preparing a more thoroughgoing deniability, perhaps to constrict the painful environmental information of recent years, scant as it has been, or perhaps in preparation for the greatly expanded plutonium extraction and waste storage and dumping that will take place at Sellafield and Dounreay in the near future.

In its straitened circumstances the NRPB has begun to accept privately contracted work, though it is said to be loath to do so because these projects take time away from basic research on the biological effects of radiation. In other words, private interests share with the government the right to set the agenda of Britain's "radiation watchdog."

At the same time, the NRPB maintains Britain's place in the nuclear councils of the world, because "Britain has to live with international standards on radiation if it is to persuade the public that it is behaving responsibly." According to the article, such standards "tend to be cooked up" by international committees. This kind of language gives a fair sense of the function and standing of such regulations. The legalism of our culture predisposes Americans to believe in their potency. But the British do not assign any significance even to standards they set for themselves, when advantage lies in ignoring them. Crown Immunity is an elegant concept and an important fact. It means that law is not considered fit to act as a restraint on government. That this notion shades off into similar protections of private economic interests is an aspect of the problem the British have always had in making distinctions between government on the one hand and economic power on the other. The extraordinary devotion of a government enterprise like British Nuclear Fuels to profit has been much encouraged by ministers of both parties. Considerations normally assumed to weigh in the thoughts of government—for example, foreign policy, an art which must pass into desuetude as trade in plutonium spreads through the world—obviously count for little over and against profit. There is absolutely nothing else to be gained. Yet profit must be overtaken fairly quickly by an array of misfortune.

This is very alarming. Members of the British Parliament are employed as paid lobbyists and are not required to declare the interests whose advantage they are paid to seek in their role as Parliamentarians. An arrangement so well suited to invite and express venality might be expected to produce it in a very pure form. Yet no mere mammonism is sufficient to account for so dismal a project as Sellafield, and what the *New Scientist* article refers to serenely as "the propensity with which British Nuclear Fuels polluted the Irish Sea with its radioactive discharges."

This piece of journalistic muddle will give a fair sense of the form in which information appears in the British press. I am extremely reluctant by now to postulate guile, simply because this strange pattern is so ubiquitous.

As often, very bad news is presented obliquely, in tones that suggest a special good fortune is being described, a government full of humanity in the form of foible and limitation. The actual news content in the piece is startling. The government disavows competence in matters nuclear but will curtail and direct the work of those who are competent to advise it. The staff of this supposedly independent agency is headed by a health physicist present at the creation of the world's dirtiest nuclear facility. Greenpeace members, described as floating offshore from Sellafield and "watching in amazement as effluent gushed out through the pipe," go afterward to the NRPB to see if they have been irradiated by a system constructed with its director's blessing. The board scientists have marketed their time to private companies and therefore cannot "expand their knowledge of the biological effects of radiation," yet they are called in to advise on "the effects of the radiation that seeps out of Sellafield on people living in the area." The board sits in on international committees to persuade its own public that "Britain" is behaving responsibly. Radon gas "pervades every home in the country."

As if to refine upon perfection, the article concludes that *Chernobyl* has brought attention to the need to study the biological effects of radiation, having only two paragraphs before alluded to the radiation exposure of the population near Sellafield. Since this radiation exposure is precisely the responsibility of John Dunster, among others, it conforms entirely to my model of the workings of the British official mind that such questions should focus on an event in the Ukraine.

John Dunster also figures very prominently in Patrick J. Sloyan's article on Sellafield in New York *Newsday*, May 20, 1986. It seems he has a map on his office wall where he can point to the "lake" of plutonium Sellafield has created off the British coast, an "almost 300-square-mile elliptical area at the end of the pipeline," the residue of the more than 500,000 curies of plutonium the plant has poured into the sea. What does John Dunster say as he ponders Lake Plutonium? " 'So damn expensive, hard to believe they throw it away.' "[*]

But this remark raises a very important question. To say that plutonium is expensive is to say it is valuable. In fact, only plutonium 239 is usable in nuclear plants or weapons. An article in *New Statesman* reports that there

---

[*] "In England, a Nuclear Plant Slowly Poisons Land and Sea," p. 1.

is no limit on releases of the unstable isotope plutonium 241, and that 550,000 curies of it have been released into the sea.[*] Are only those emissions limited or measured that represent straightforward economic loss? By "plutonium" do Dunster and others mean only plutonium 239?

John Dunster can take heart, however, if he is oppressed by the loss of plutonium. Walter Marshall, who looks in photographs like Tweedledum, but who has been made a lord for his previous accomplishments and is head of the Central Electricity Generating Board, has found plutonium in nature, surely the crown of his career. Recently he has given lectures in which he reassures the public that the small amount of plutonium released in a recent, much ballyhooed accident was only equivalent to the plutonium "naturally present in the top yard of soil over an area of just five square miles." This perspective was admiringly reported by James Wilkinson, BBC-TV science correspondent.[†] A streak has been added to the tulip. It seems to be this man's happy genius to bring perspective to nuclear issues. He is reported to have won applause at a meeting of the British Nuclear Forum by saying that the effects of radiation exposure within the Chernobyl exclusion zone would be "no worse than smoking a couple of extra cigarettes a year."[‡]

*   *   *

---

[*] "Fitting in with the locals," September 14, 1984, p. 9.

[†] "How safe is Sellafield? Claims about nuclear leaks should be handled with care," *The Listener,* February 27, 1986, pp. 2–4.

[‡] "The Chernobyl syndrome: The day the impossible happened," *The Observer,* May 4, 1986, p. 9.

Strange as it seems, the explosion at Chernobyl has been turned to great advantage, and with worrisome ease. That such a major event should seem to have had such limited effect is used to cast doubt on the legitimacy of all anxieties about man-made radioactivity in the environment. A recent essay in *The Observer* launches off with an attack on the press by a Cumbrian man named John Allan, comparing reports which described more loss of life at Chernobyl than Russian authorities subsequently confirmed, to reports about Sellafield—also, according to Mr. Allan, lies meant to sell newspapers.[*] Typically, the article compares the dangers of working in a nuclear plant to those of coal mining, and notes that closings of mines and steelworks have made the area overwhelmingly dependent on Sellafield in any case. Then it establishes that plutonium contamination incidents, called by the workers "taking a bit of ploot," are entirely commonplace. Dr. Jack Strain, head of the medical staff at Sellafield, told a reporter that if they informed doctors whenever their patients among the work force were contaminated, "we would be writing 100 letters a day." A man named Jim Horspool, apparently involved with Sellafield from its early days, propagates the view "that a certain amount of radiation could be beneficial." As such people always do, he adduces the fact that we are exposed to radiation in nature. " 'Chadung,' he cried. 'That was a cosmic ray going straight through. *Chadung.* There goes another one.' " Then we are introduced to "Atomic Stan," the worker contaminated in 1957, when he was among those who put fire hoses into the burning Windscale reactor. He describes having looked into the core itself. The son of a stone-

---

[*] "Sellafield spurns the bunker image," May 25, 1986, p. 52.

mason who died in middle age of stone dust in his lungs, Stan is now seventy-one and smoking thirty cigarettes a day. The article describes tours of the plant arranged by BNF, and an exhibition which demonstrates the radio-activity of Cornish granite and of a luminous watch face.

The gist of the article is that radiation has indeed received a bad press. Workers accept the risk with "a kind of quiet pride," a risk that compares favorably with those of other industries, or at least bears comparison flattering to both traditional industry and the nuclear industry. The quiet pride of Sellafield workers, we are told, is like the attitude of miners toward the risks of lung disease or cave-ins. This is a disturbing evolution in industry apologetics, since it was supposedly the benev-olent hope of the nuclear industrialists to relegate such suffering to the brutal past.

There are stories in the press which give insight into the obscurity surrounding the health issue, for example, one in the *The* (London) *Times*[*] about a man named Harry King, who worked inside the Sellafield plant in a room with an inoperative air filter. He was found to have been exposed to an overdose of plutonium. In the course of time his teeth and hair fell out. He developed cataracts, and finally died of brain cancer. BNF paid his widow £8,000 compensation but did not accept re-sponsibility for his death. The physician at Sellafield, Dr. Jack Strain, has said that when workers are contam-inated it is explained to them that no one has ever devel-oped side effects from working in the nuclear industry. This assurance might have required some revision after Chernobyl, though perhaps not, given its fairly spectac-ular imperviousness to deaths like Mr. King's. BNF, as

---

[*] "Surviving in the nuclear shadow," Sally Brompton, February 28, 1986, p. 15.

of February 28, 1986, the date of the *Times* article, had paid £246,233 in compensation to Sellafield widows. This includes an award of £120,633 in 1985 to the family of a member of management staff.* The company is initiating morbidity payments to those of its workers who are in the way of succumbing—to *Weltschmerz*, presumably.

These attentions do not fall like the gentle rain from heaven. BNF holds hearings at which claims for compensation are accepted or denied, on what basis God knows, unless it is the poignancy of the tale. If no one dies from working in a nuclear plant, obviously no one can be more or less deserving of compensation on other than sentimental grounds. Certainly the anguish of dying from nameless causes is much intensified by developing cataracts and losing one's teeth and hair. So there would exist a humanitarian basis for such selectivity. And BNF has made cancer its favorite charity, and is therefore no doubt particularly moved when this malady appears among its work force.

Most cases are denied hearing in the first place—94 out of 164, according to the article in *The Times*, which explains that BNF had set up its compensation system to avoid the expense and delay of legal procedings. This is an impressive tribute to the government's faith in their objectivity, more striking in view of their stated lack of confidence in the plant's management during this same period. The figures suggest settlements in sixty-eight cases averaging less than £300, or about $500. An indication of the job status of those who do most of the dying, presumably. Thrift, at least, appears to be served by the arrangement.

---

* "£120,633 award after radiation-link death," *The Times*, March 27, 1985, p. 2.

The importance of the assertion that no one dies from working in a nuclear plant to those responsible for the management of Sellafield would seem to militate against pure objectivity in adjudicating the grievances of those who feel a family member has in fact died from working in the nuclear industry. To admit to responsibility in any death would establish that there are conditions in which exposure to radioactive materials traumatizes the human organism fatally. The ramifications of such a concession would be very great, the uproar among widows and orphans being only a leitmotif. For it is the impeccable safety record of the industry itself as much as anything that predisposes the government to view with skepticism any suggestion that they might be creating health problems in the general public. In fact, comb the literature as I may, I find no allusion to experiments of any kind that bear out the view of the authorities about the nature of radioactivity. Rather than producing research that indicates the degree of innocuousness of radioactive materials required to justify practices at Sellafield, they defend themselves with claims that their harmfulness cannot be proven. "An observed association between two factors does not prove a causal relationship," in the words of the Black Report on childhood leukemia near Sellafield.

There are many occasions in which it seems that the interests of industry must influence scientific research. The Black Report, granting elevated levels of both radiation and leukemia in the area of Sellafield, and observing that irradiation is the only known cause of this illness, recommended research to discover other possible causes. Subsequently the Human Leukemia Virus Centre has been founded in Glasgow to research the role of viruses in the onset of leukemia, specifically in the phenomenon of leukemia clusters.* Such excesses within geographical areas have been found near many British nuclear sites.

Since both leukemia and irradiation depress the immune system, I suppose we will all be kept busy reading about viral infections associated with cancer. It will be implied, I am sure, that observed association should be construed as causal relationship.

An article published in the *New Scientist* in 1986 titled "Clampdown at Sellafield" describes a bold and assertive government that has "decided to step in and curb airborne radioactive pollution from nuclear plants," startling both BNF and environmentalists "by announcing plans to set both an overall limit and numerical limits for specified radionuclides and groups of radionuclides."† Interestingly, this initiative was taken just at the same time the cloud was floating west from Chernobyl, and thus, presumably, by those very same gentlemen so unschooled in such matters as to have boggled at big words like "isotope." This little article goes on to explain: "At present, airborne emissions are monitored—but there are no specific authorisations, as there are for liquid discharges, to control permitted levels of gases, mists or dusts containing radioactive contamination." However, the article notes serenely, "the company is worried that it may have problems keeping to the new limits if it wants to maintain the throughput of its reprocessing activities while vital pollution control plant is out of action during maintenance." In other words, no variety of emission is controlled at all except what comes down the pipeline,

---

* "N-plants may be in clear over cancers," *The Times*, December 9, 1988, p. 6.
† May 8, 1986, p. 22.

and the priority given to "throughput" is such that the plant continues to function even while "vital pollution control plant," whatever that can amount to, is out of commission. Once again, it is surely reasonable to wonder what all the NRPB's research was for, if these emissions have been unlimited all these years.

This is no Faustus legend. Science is not at the root of the problem, nor is an overweaning desire for knowledge, or perhaps even power. The plutonium factory at Sellafield merely thrives on the failures of science, as an undertaker thrives on the failures of medicine. Any great curiosity about the mysteries of Being would have led these gentlemen to read up on their stock-in-trade—a pardonable hubris, surely, beside the incurious good cheer with which they have scattered it over the landscape. The refuse collector in Brazil who stripped the lead shielding off the capsule of cesium 137 to sell it was engaging in a naïve commerce, very modest in scale but similar in kind to the commerce centered in Sellafield. He did not know what he was doing. It is as if the refuse collectors at Sellafield do not know what they are doing, either. Cesium 137 is among the contaminants released in important quantities through their pipeline into the sea, and among those toxins most present in seafood. The terrible deaths of the people in Brazil who came in contact with this substance indicate that harm is to be feared from exposure to it. Yet not until the accident at Chernobyl, when it fell over Europe and was treated, briefly despite its persistence, as a public health problem on the Continent, was it treated as a problem in Britain, where it has accumulated in the environment for decades.

But can I suggest that men who have long mingled with the premier nuclear scientists and technicians of the world do not know what they are doing when they pour radioactive materials into the European environment? Is another conclusion possible? This is not a rhetorical question. While behavior of the kind I describe is what one fears of terrorists, and while there can only be malice at the bottom of policies so abusive that an enemy might blanch at the thought of occasioning such everlasting, indiscriminate harm, still is it to be imagined that these men have quite knowingly set out to do what they have done?

They must have known better once.

A London *Times* supplement devoted to Calder Hall, on the occasion of its commencing operation in October 1956, included an essay titled "How the reactors were planned," written by B. L. Goodlet, engineer in charge of design study, which described radiation hazards as follows:

> The term contamination implies loose radioactivity—dust or droplets—which may be absorbed into the body through skin abrasions or by breathing, eating or drinking. The counter-measures against contamination are complete enclosure of all processes involving radioactive materials and rigorous controls of the effluent—gaseous, liquid, and solid—from all plants using radioactive substances.[*]

The plants were not designed to meet this standard, however, and over the years theory seems to have conformed itself to practice, becoming primitive and improvised to suit the needs of the industry.

I think moral aphasia might be a useful concept. It

---

[*] P. v.

is no doubt apparent from my long approach to Sellafield that in my view a civilization with such a pervasively violent history, in the course of which it has acquired the highest estimation of its own decency and mildness, has developed a peculiar trick of mind, not to be called a divided nature, since the conviction of particular goodness always one way or another justifies or conceals or expedites really remarkable transgression.

Any schoolchild knows better than to do what these men do, transporting toxic materials over thousands of miles of sea despite the risks of accident or seizure—do we know these things have never happened?—or bringing them across the Channel, and through London by train, and then storing them in quantity among leaking silos and earth ditches, until they can be bathed in solvents, the greater part of the uranium 235 and plutonium 239 extracted from the resulting broth, and the rest flushed into the sea. Then plutonium 239—bomb-grade, that is—is stored at Sellafield, or returned by sea or by air to whomever it was that sent it to Britain as waste, so that it can abide forever as plutonium in undisclosed quantities in unknown hands in a world not remarkable for stability or, as I have just demonstrated, good sense.

In 1976 then-vice-chairman of BNF, Julian Avery, was reported by *The* (London) *Times* to have answered concerns about the possibility of terrorist actions in a "plutonium economy" with the remark that it is not plutonium but terrorism that should be eliminated.[*] Presumably Northern Ireland demonstrates the ease with which this is to be accomplished.

---

[*] "Mr. Benn rules out need for quick decision over fast nuclear reactor programme," December 14, 1976, p. 4.

*   *   *

Like so many of the world's sorrows, Sellafield, then Windscale, grew directly out of World War II. The United States attempted, after the success of the Manhattan Project, to monopolize nuclear technology, excluding the British along with everyone else despite the early prominence of British physicists in atomic research and despite their contributions to the building of the bomb. Newspapers and magazines of the period make it clear that relations between Britain and America after the war were not especially cordial. A member of Congress declared himself no more pro-British than pro-Russian, in response to Churchill's Iron Curtain speech, delivered in Missouri, which raised hackles among the American public by seeming to propose an Anglo-American alliance.

The closing down of access to technology seems to have been a reflex of pure alarm by a government that had arrived rapidly, though still too late, at a sense of the perils of "proliferation." Since the United States never had any sort of corner on the relevant science—British accounts of the evolution of nuclear physics hardly include an American name—there is little reason to doubt that the attempt to control it was futile from the beginning. Margaret Gowing, historian of the British nuclear enterprise, declares that the bomb was their invention. I would eagerly concede them that distinction.

In any case, the British received no special treatment from us in the matter of atomic weapons development. As a result, they set about developing them on their own, with no other resources than "green fields and grey matter," in a phrase of the period, and with such urgency that the construction of the plant at Sellafield was, as I have said, supposedly compromised by materials shortages brought on by the war. Since this fact is now used to account for the plant's spectacular history of accidents, no major reconstruction of the basic plant can have been undertaken when time and the availability of materials allowed. Whether the jerrybuilding of the site was indeed a product of haste or not, the haste is notable in itself. The possession of atomic weapons established nations as powers, and Britain, still an empire at the time, was determined to keep its place in the counsels of the mighty. The United States may have been pressured by a fear that the Germans would develop the bomb first, and Russia may have been driven by the fear of conceding such an overwhelming advantage to a possibly hostile American government, but the British seem to have understood earliest the status-conferring properties of these new weapons. Their haste had no justification in terms of any threat to their safety. It must be laid to pique, and to a desire to retain their importance in the world. They had made a fresh demonstration of the practical value of their empire in the war just concluded, summoning troops from the ends of the earth. But Britain (perhaps England would be better here) seems always to have felt so vulnerable, in the matter of its wealth and importance, that any course of action or any policy could be justified as the expedient of desperation. Though Britain enjoyed greater relative wealth and dominance for a longer period of time than any other power after Rome, its behavior has always been predicated on littleness. At the end of World War II Britain had by far the strongest economy in Europe, and banks full of savings. There was no more pressing need for cheap than for hasty nuclear development. Yet both are treated now not only as justified in themselves but as excusing their inevitable consequences. A great continuing belief in the wisdom of early weapons policy is reflected in the ability of the Labour Party to destroy any electoral advantage that seems about to accrue to it by proposing that Britain rid itself of nuclear weapons. The party seems now headed for marginalization on the unaided strength of this issue, even while its constituency has a great practical need for a functioning party—a need which, in fairness, Labour is poorly suited to fill in any case, less so since Mrs. Thatcher abolished the elected major city governments that were Labour's power base.

Britain has never had to justify to its people the possession of nuclear weapons on any grounds other than their assuring that the country would continue to cut an impressive figure on the world stage. It has never had to excuse the slovenliness with which the great enterprise was gone about on any grounds other than Britain's littleness and poverty. Presumably it is the country's exceptional depth of civilization, or its gift for taking the long view, that elevates it above the level of *poseur* in the minds of its people. In any case, it is assumed that Britain should be a world power, that extraordinary methods are necessary to make it one, that developing nuclear weapons was an appropriate course to take in assuring Britain's place among nations. While we and the Russians pardoned ourselves as the defenders of opposed systems which we were persuaded could bring justice and the alleviation of suffering to other societies and generations, Britain, typically, scrabbled to shore up its own interests, narrowly defined. Seldom departing in public from the values with which

Americans identify themselves, except to add emphasis in one instance, urge temperance in another, and imply, through a certain fretfulness, that the whole thing would be better if it were not so crudely managed, the British nevertheless acquired nuclear weapons to establish their independence from the United States, in heat and haste that look like desperation. Denied the use of testing sites in America, they considered exploding atomic bombs in Scotland and Yorkshire, and finally arranged to use Maralinga in the Australian interior, which they left severely contaminated with plutonium, as it remains to this day, as well as islands off the Australian coast, sending one especially notorious cloud drifting across the mainland from Monte Bello. Now, of course, the British test in Nevada.

The materials for these handsome acts of national self-assertion have come from Windscale/Sellafield and from Calder Hall, the plutonium-producing, gas-cooled, graphite-moderated reactor at the same site. As a plant primarily designed for the production of plutonium, with electricity generation as an ancillary function, Calder Hall served as prototype for the first generation of nuclear reactors in Britain. There has never really been a second generation. Not only were military and civilian uses not distinct, the former took priority over the latter in determining reactor design for both uses, since the reactors are not especially efficient as producers of electricity. Dounreay, in Scotland, was launched very early, in 1959—a breeder reactor, the wave of the future, or so it was thought at the dawn of the nuclear age, and so it may prove to be, insofar as these developments have not precluded a future. Plutonium would be needed to fuel such plants when the technology was perfected. Perhaps the British decision makers hoped to be the first into the field, with a prototype for the breeder and a supply of the material to run it. Export sales have always been a consideration of the first importance in the development of nuclear power.

Britain's history especially predisposes it to keeping a world market in view, so perhaps such an explanation would account for the otherwise strange decision to build so many plutonium-producing plants, more than have ever been required to supply the fantastic demands of nuclear weapons development in the United States, for example. British strategic nuclear weapons seem to be numbered, for the purposes of international reckoning. Since even the British Parliament has no access to defense information, and since governments can conceal nuclear capabilities as they choose, the seeming disproportion between the scale of Britain's plutonium production and its apparently modest defense needs is not based on information of sufficient quality to merit brooding over. Understating numbers of warheads would reduce pressure on Britain to enter into arms control negotiations, while overstating them would enhance national prestige at little cost and little risk, the force being meant to trigger grander events, not to sustain an exchange on its own.

The relation between British plutonium production and weapons production is asserted as often as it is convenient to invoke security and the national interest in behalf of Sellafield, though the plant has for a long time functioned mainly as a repository for foreign waste. A Martian, watching Germany and Japan ship toxins into Britain to be poured into its environment, might wonder why so much blood and sweat should have been expended in the defense of this same island.

An occupying army is after all a survivable problem, and democracy a nostalgia where a government can, in secret, put an ax to the root of the culture that supposedly sustains it.

Be that as it may, Sellafield is a significant part of British national defense, as press and government reckon these things. Sellafield may indeed epitomize the phenomenon of national defense in the nuclear age, being a vulnerability not to be dreamed of in any country less well defended, a vast bull's-eye for enemies or terrorists or plain misfortune.

Affronted by the American attempt to keep a monopoly on nuclear technology, the British set out to build bombs. That is how they tell the story. But if the technology had been handed to them, presumably they would have used it to build bombs. Why else should their exclusion have aroused such shock and frustration? In any case, they put a complex together to produce plutonium, tested ferociously where the locals permitted, built an entire series of plutonium-producing reactors with electricity production as a secondary feature, reprocessed and stocked plutonium, and built Dounreay, which has not functioned properly since it was switched on; which has a reprocessing plant of its own, sending radioactive effluents into Scandinavia; and which nevertheless sends wastes to Sellafield to be reprocessed, by stormy sea.

The output and commerce reflected in all this are clearly of startling magnitude. What Sellafield reflects in terms of the intentions of those who set all this in motion is probably not a simple question. The more secretive and narrowly based decision making is, the more eccentric it becomes, and often things happen for reasons that are foolish or bizarre, and therefore elude all surmise.

Plutonium is an extraordinary substance, both waste and commodity, costly and dangerous to keep, dangerous and profitable to sell. Where it exists in quantity, as it does in Britain because of its Magnox reactors, and more recently because its expanding of reprocessing services brings in the plutonium-rich wastes of a great part of the world, there would be considerable pressure to find ways to use or be rid of the stuff. Ninety percent of the nuclear waste that has been dumped into the sea has been put there by Britain, presumably as a function of its role as universal dustman.

I speculate that the origins of it all may lie in that first decision to produce nuclear weapons, which led to the development of a plutonium-producing reactor, which was subsequently sold abroad as a power-generating reactor. A notable feature of the British Magnox reactor is that it produces wastes which cannot be stored for long periods. Reprocessing is a solution to this problem.

National prestige as well as enormous sums of money are at stake in all nuclear transactions. When Chernobyl exploded, supplanting the fire at Windscale as the most serious nuclear plant accident in history, Britain is reported to have brought pressure to bear on Italy to prevent it from closing a Magnox reactor; that is, one of the type of Calder Hall and the other British reactors of the first generation, and similar to the one at Chernobyl.* (British and Soviet physicists worked together in the pioneering days, so the similarity of their industries may suggest intellectual cross-pollination.) The economic interest of Britain as an exporter of nuclear technology, and also the interest of the British government in avoiding anxiety at home about the

---

* *Chernobyl: The End of the Nuclear Dream,* by Nigel Hawkes, et al., Vintage Books, 1987, p. 153.

safety of these old reactors, seems to have prevailed over Italy's disquiet.

Italy is among the countries whose waste is reprocessed at Sellafield. A benefit of selling reactors abroad which produce unstable wastes might be that the buyer will also be obliged to pay for the disposal of spent fuel rods. The potentialities for turning a profit are considerable, given the will. The history of Sellafield certainly demonstrates the injustice of Mrs. Thatcher's chiding her countrymen for being laggard in this regard, by the way. But then it is always the case that people find themselves deficient in the things that in fact matter most to them, or seem to them most admirable, and in which they are least liable to be remiss. The British berate themselves with excessive caution, and with being slow about putting scientific discoveries into commercial application. Any student of the history of Sellafield will surely find them innocent on both counts.

It is clear from articles in the press of the time that the first British reactors were not considered an optimum design, even by those responsible for building them. That they should be gas-cooled was a choice encouraged by thrift. Since the old plants continue to function, long after the end of their design life, their deficiencies are by no means matters of academic interest, any more than their characteristic of producing "waste" with a high plutonium content.

Building reactors whose waste could not be stored locked Britain into reprocessing continuously and on a large scale. Sellafield, which was built to make fissile materials for bombs, has assumed the civil function of waste dumping, in the course of producing the same fissile materials. It has made an officious and energetic show of managing the unmanageable. Its patrons or clients in other governments have in fact done the same thing. They have evaded the most important and costly problem created by nuclear energy by paying the British to take if off their hands, temporarily, since winds and currents assure that their problems will not remain solved for very long. In the meantime, they can appear to have mastered this most difficult technology, and they can propagate their versions of it through the world. The clientele of Sellafield is a *Who's Who* of technologically advanced countries: Japan, Italy, Germany, Switzerland, Spain, Holland, and Sweden. France has its own pipeline into the sea at Cap de la Hague on the English Channel.

The relation between the civil and military functions of Sellafield is never clearly defined—a fact which limits its accessibility to EEC inspection, among other advantages. One Christopher Hinton engineered the pipeline and received an honor for it from that same government which forbade him to mention it in public. This discretion is consistent with anxieties about the wisdom of dumping toxins into the sea. Yet there is no indication that concepts like cumulative impact or long-term consequences were brought to bear in making the decision, though Britain was embarking on a policy whose course would be measured in decades at the least. The silence about the pipeline must indicate lingering doubt, or the anticipation of criticism of the kind that came up at the United Nations in 1958, when John Dunster defended the dumping as an experiment. While the results of this "experiment" revealed contamination—though at levels acceptable to the government—a more characteristic defense of Sellafield asserts that plutonium, being almost as heavy as lead, will lie on the sea floor, presumably somehow inert. Water passing through lead pipes is contaminated. So it seems logical that currents and tides passing across a sea floor on which rests a pool of sludge, including plutonium ash, would also become contaminated. In neither case does weight impede the process. Here again it should be borne in mind that plutonium has never been the sole or primary radioactive material released into the environment. So its peculiarities could not in any case justify the emissions from Sellafield.

While I am no physicist, I do know that radioactivity was observed in nature in the first place because certain substances give off small amounts of heat. Plutonium, and especially americium, into which it is converted as it decays, give off heat. Modest as my experience of the natural world might be, I am bold enough to suggest that when heat-generating materials are spread across the floor of a northern sea, a new influence on the movement of water is introduced. In Britain, even at this hour, learned men are struggling to account for the ferocious toxicity of spume in this unhappy region. I believe my hypothesis might be of use. Since contaminated water would be warm, it would come to the surface. The contaminants, being warm, and continuously warming the water around them, would not sink again. At the same time, more contaminants would be added from below. So the model according to which substances which are carried into the sea in water suddenly become impervious to being dispersed by water once they are in the sea seems flawed on several grounds.

Of course I am proceeding backward from observed phenomena, such as radioactive spume, radioactive sand, and radioactive wind. But this method is certainly superior to insisting on the appropriateness of a model which runs counter to observed phenomena. An amateur's interest in the virulence of spume might distract attention from what really is a more important question—that is, why, with plutonium and all the rest pouring into the environment daily, it should be important to determine by what process the surf has become especially hot. The source of the problem is not far to seek, and the nearest way to containing it is equally clear. While dumping continues at an accelerating rate, with no prospect except for invidious change, to trace the movement of radioactivity from the bottom of the water to the top seems a misappropriation of effort. It is of a kind with the sponsorship of cancer research by British Nuclear Fuels, in that it tends to distract attention away from a situation that is extremely straightforward, toward its intractable consequences. Research projects and furrowed brows and a fluttering of white coats, things full of reassuring implications, at the same time create an aura of mystery where all is as plain as day.

The British government has just decided to build a pressurized water reactor at Sizewell, in England, after an inquiry that seems to have satisfied notions of rigor by continuing for months and producing, literally, scores of tons of testimony, and to have satisfied definitions of civility by ending with funny-hat party. As always with such inquiries, the government set the terms of the question and appointed the inquiry chairman, who not only presided over the affair but also determined its outcome—which is not binding on the government. Every bet is covered. The issue at Sizewell was whether the plant to be built was to be an advanced gas-cooled reactor, a British design, or a Westinghousetype pressurized water reactor (PWR). Environmental-ists were allowed to testify, though what they said against nuclear power itself was necessarily extraneous, and they appeared, when the time came, funnily hatted. The famous weight—that is, heft—of the information gathered in this inquiry certainly reflects the fact that elaborate cases can be made against the construction of reactors of either kind. Yet it is cited as evidence of the great deliberation with which Britain approaches nu-clear decision making.

The committee chairman, Frank Layfield, decided in favor of the PWR, a decision that confirmed the wisdom of the government, which had favored it from the first. The decision was supported with assurances that the American design would be brought up to British safety standards. This may not involve great expense. Sizewell is already the center of a leukemia "cluster." Both Sellafield and Dounreay satisfy British safety standards. Less extraordinary plants must therefore fall within them quite effortlessly.

In 1983 it seems a spill occurred at Sellafield which caused the beaches to be closed for months. This was one of those clustered events which must make the management at Sellafield, together with the British government and any lover of mankind, say alas and alack. In my credulous days I considered this an authen-tic and significant event, but I have begun to realize it bears scrutiny badly. I was living in England that year, and not long before the spill was reported, as I perused *The Guardian*, I came upon a little article on a back page in which it was stated that experts had begun to believe that it was more harmful to ingest plutonium than had previously been thought. Of course I was startled. For just then American public opinion was turning over as if for the first time the fact that we and our great competitor were acting as impresarios for the Day of Doom. Of all the terrors we had prepared for ourselves, nothing compared with plutonium, according to the lore shared at dinner parties among the thoughtful classes. The word "plutonium" leaped at me, conditioned as I was. I began to notice mention of this substance and its ilk in the news almost daily.

One accepts the news, for some reason. A fiction writer has to braid events into a plausible sequence. But the news is simply a series of reported incidents which, one assumes, manifest varieties of accident and causation, plausible if they were known. There are no grounds for this assumption. Sometimes the news reads suspiciously like unusually clumsy fiction. In litigious America, with its habit of trials and investigations, we always attempt to establish a narrative on the order of who did what to whom and why. This approach is full of problems, the chief one being that people become loyal to one narrative or another and lose interest in objectivity, together with respect for information that fails to confirm a favored version of plot and character. I can only assume many Americans read the same articles I read. They swarm that island, to the utter weariness of the natives and one another. But England is established, in their narrative, as a mild and scrupulous old nation which, like the lion in Mark Twain's unfallen Eden, gorges on strawberries. So they see and do not perceive, hear and do not understand, full of that awe which

Toqueville, precisely wrong, said Americans were incapable of feeling.

According to our narrative, poor Europe must endure our crude embrace or fall into one more crushing. It is a funny story, really. All the troglodyte behavior for which we can never apologize abjectly enough, all the stationing of troops and sticking the world full of missiles, are the single bond that ties us to Europe, the one proof, never sufficient, of loyalty and love. We make ourselves the worthy object of our own contempt for the sake of a civilization much richer and more populous than the Soviet Union.

We cannot absorb or retain information which would establish Europe as a major and not at all a brilliant actor in contemporary history. We have sacrificed our humanity to preserve countries that connive in the production of the worst sort of explosives and toxins and their release into the environment. There is nothing pleasant in this fact. It has no place in the story. When the Russians hear all the prattle about Western values they must surely assume they permit plutonium dumping. Who can argue? It may be more than their absolutist history that prevents them from being converted by the force of Western example.

I often wonder how the Russians interpret the silences and omissions in American journalism. Every country is ridiculous in its own way. We are at our most ridiculous when we imagine our bold, hectoring press with its armor of legal protections informing us, in some meaningful sense. Any grumbling about its excesses is shamed to silence by a reminder of the great freedom the press must have in order to do its great work. It is impossible to know the extent of its misfeasance without leaving the country. When Americans speak of its failures, they usually impute them to public indifference to foreign affairs, as if fine, plump articles on these subjects withered on the shelves, as if the American consumer pinched an item about former Swedish Prime Minister Olaf Palme's secretly developing nuclear weapons over an eighteen-year period, and squeezed an item about the British secret service raiding the offices of the *New Statesman* and the BBC, and rejected them both in favor of an update on the Betty Ford clinic. News of the affairs of the world is not readily available to Americans. I believe this may be true for Russians and Chinese. If the origins of isolation are different in the three countries, that is less important than the fact of isolation, which is equally incompatible with democracy in every case.

The American zeal for establishing a narrative context for events may falsify as much as it clarifies. But it does at least set events one beside another to see how they cohere, and it acknowledges the importance of actors, who are assumed to be responsible and accountable for what they do. British news, by comparison, is simply a series of revelations. For thirty years a pool of plutonium has been forming off the English coast. The tide is highly radioactive and will become more so. The government inspects the plant and approves the emissions from it. The government considers the plant poorly maintained and managed, and is bringing pressure to lower emissions. The government is expanding the plant and developing another one in Scotland. Foreign wastes enter the country at Dover and are transported by rail through London. Finished plutonium will be shipped from Scotland into Europe by air. Whose judgment and what reasoning lie behind these practices and arrangements? The question is never broached.

Information merely accumulates, without effect. The British government, the great constant behind the notional shifts of management, the proprietor and stockholder, never loses its ability to reassure the public, assuming the lofty role of inquirer into its own doings and finding nothing seriously amiss, nothing a little finger wagging will not put right, a little expression of lack of confidence in the management. It is very much as if the object of these revelations was to let the public know what it must accept.

Certainly the most striking effect of all the revelations to this point has been to produce quiet, while the government launches into the vast program of construction that will make Britain an ever greater center of plutonium extraction and waste dumping. Some of the writers and publications from which I have taken information seem courageous. But then since none of these articles seems to have done more than to inoculate public opinion against bad news still to come, how is a foreigner to judge any writer's intent? No hearing will convene to assess the wisdom of shipping radioactive wastes through a populous capital, or dumping them into the sea, or extracting weapons materials from them to be shipped by air into Europe, and through North America to Japan. Bad news only intensifies the prevailing resignation.

I noticed the little article about possible adverse effects of ingesting plutonium. It was feared, the article said, that children absorbed the material many times more readily than adults. Soon afterward the matter of the contamination of the beaches of Cumbria arose. First an employee of the plant, nameless and faceless as figures in this narrative very often are, stopped to tell a young family not to allow their children to play on the sand. They wrote to their Member of Parliament, who raised a question about conditions near the plant. At about this time, a Greenpeace boat went out to cap the pipeline by way of protesting the gush of radioactive materials and solvents into the weary sea.

This is a very strange little story by itself. Something over a million gallons go down that pipeline in the course of a day. Could people working under water actually hope to cap a double pipeline through which so much toxic liquid was flowing? Capping the pipeline at Sellafield, if it could be done, would seem to involve the risk of backing up this enormous outflow, flooding the beach or the interior of the plant, a dubious piece of environmentalism. Their ability to close the pipe was said to have been taken seriously by someone and foiled. The mouth had been changed so that the cap they had prepared for it would not fit, a fact that led to speculation in the press about government surveillance of Greenpeace.

As the event was reported, these Greenpeace divers first went into the sea at the mouth of the pipeline, to take silt samples. They surfaced again through a slick, and discovered, when a Geiger counter in the boat indicated radioactivity at 1,500 times "normal" levels, that they were contaminated. Thus was discovered the great slick that closed the beaches of Cumbria, that made them get up and move, like Birnam Wood. (I have never seen a photograph of this giant convoy of trucks, moving back and forth, presumably for weeks, so I cannot speculate on the methods used to avoid spreading contamination en route to the site of disposal,

wherever that may have been. Plant management denied the beaches were being removed. A large decontamination effort was undertaken, however, and since
flotsam and seaweed in themselves could hardly have
been sufficient to sustain such an effort, I incline to
believe the reports that sand itself was removed. Testimony at the trial of BNF for its management of this
incident described levels of radiation at ten thousand
times background.* While it is not possible to assign any
precise meaning to such figures, they clearly indicate an
extreme situation. The trucks disappear from later
accounts altogether. So does the attempt to cap the
pipeline.)

Some days passed between the contamination of the
Greenpeace divers and the closing of the beaches, which
seems to have been the time it took for BNF to acknowl-
edge that anything unusual had occurred. It has never
been clear to me whether they did not know a spill had
occurred, or whether they did not consider the event
unusual. In all such events, delayed response is characteristic.

If the spill was serious enough to require a significant
effort of decontamination, it would certainly have been
prudent as well to relocate children and pregnant
women while the decontamination was proceeding.
When I imagine these residues of spent fuel, oxides fine
enough to be carried in solvents into the sea and then to
be brought in again by the tides and winds, I can only
imagine that they would be highly particulate, and that
when the sand was disturbed they would be winnowed
out by any movement of the air, unless the sand was wet,
in which case they would seep out with the moisture. In

---

* "Sea waste 'radioactive oily slick,' " *The Times*, June 21, 1985, p. 5.

other words, I cannot imagine how the repair of the beaches could have failed to have intensified the contamination of the area, in terms of unavoidable human contact with it.

I find the going a little hard when I try to imagine a boat full of bright young men, literate in matters nuclear, with a Geiger counter on board, on their way to take silt from the floor of this notorious sea. Why did they go on this mission? Because radioactive wastes were being disgorged into the sea, at that very site. Did they expect to be contaminated, diving down to the plutonium-spewing orifice? Clearly they did not. It was supposedly an oily slick that made them radioactive when they returned to the boat. Where would this slick have come from? That pipeline. Therefore, they no doubt dived as well as surfaced through it. So here is the problem. Why would fit young men with their lives before them, diving near the pipeline *because* it released radioactivity, and who had a Geiger counter along, *not* test the condition of the water before they entered it? Putting aside the apparent fact of one particular slick, how can it have come as a surprise to them that they were contaminated, and why should they have treated the discovery that they were as meaning something exceptional had happened? If they really thought this radioactive emission problem was only of such magnitude that one could dive into the thick of the most prolonged and intense contamination in the world and rise out of it as fresh as Wordsworth's Proteus, then they might make more profitable use of their time selling toy seals—the kind most resistant to radiation in the environment, as these conservation-minded folk are no doubt aware.

In fairness, Greenpeace seems to have a Geiger

counter problem. Here we read how they had one along in the boat. But then when Chernobyl blew, the only Geiger counter that was used to provide readings on levels of radioactivity on the west coast of Britain belonged to a high-school science class. Surely we might have expected a flood of independent information from this sun-bronzed band of nuclear foes, since they do have Geiger counters, as this story proves. Yet they seem not to use them to maximum effect, and that is a pity, all the more so because their shortcomings in this regard replicate precisely those omissions of government, industry, the regulatory agencies, and the scientific community which create the aura of mystery around Sellafield, an uncertainty a little monitoring could so quickly dispel.

But let us return to the matter of narrative. Let us suppose the facts discussed so far were to be construed for the uses of fiction, and the writer was obliged to impose on them or discover in them that order of reasonableness and plausibility which could keep the reader from flinging the book out the window. Clearly the nuclear activists could not leap into the most radioactive sea in the world at the eye of its contamination and then register amazement that their Geiger counter was agitated. They might at most sail out and sail back in again. The idea of capping a pipeline from which comes a massive flow of toxic materials clearly must be scrapped on grounds of implausibility. And the detail concerning the contamination of the divers and their boat had best be crossed out, too, since the reader would wonder about the other ships in the Irish Sea that day and the catches pulled up through the toxic film and stowed in contaminated hulls and carried away into ports and countries where the name of Sellafield is

never heard—America, for example. These environmentalists would no doubt be expected to think in larger
terms than merely their own persons and property.
According to reported testimony at the pollution trial—
the first in British nuclear history—there were people
on the beaches that day and fishing boats off the coast.*

Without offending the reader unduly, the tale could
be told in this way: Greenpeace declares the coast to be
contaminated, couching the information, so as to make
this fact seem surprising, in terms of the pipeline
anecdote. After a few days, BNF admits that there has
been an accident. In the fiction, this delay would make
the spill seem less alarming and egregious than it would
if the management responded with any kind of haste. At
last they admit that solvents were indeed accidentally
flushed into the sea. The beaches are closed, hauled
away, replaced, and declared safe, though strollers are
advised not to pick up bits of flotsam, which are still hot.
Except for the removal of the beaches, no extraordinary
measures are taken to protect anyone. We know from
experience with conventional oil slicks how they devas
tate coasts and aquatic life. This slick, we are to believe,
drifts in and is blotted up and hauled away—to some
corner of that vast kingdom where radioactivity can do
no harm, or back to the sea, to resume life as a slick.
Obviously the beach must be considered radioactive
over the long term, or there would be no point in
moving it. So wherever it was put, it will be radioactive,
over the long term.

This makes narrative sense, if the point and object of
it all is the removal of severe radioactive contamination
from the area of Sellafield. It would be very easy to

---

* "BNF pollution trial starts," Pearce Wright, *The Times*, June 6, 1985, p. 3.

imagine reasons for doing this. Putting aside the matter
of any particular episode of contamination, the beaches
had absorbed contaminants for years. The great concentration of radioactivity in surf and in seaweed and
flotsam would make this inevitable. At some point it
would have to become a problem, more especially
because the area is the site of an enormous construction
project which will continue over years of time. (Former
BNF chairman Con Allday has written that instruments
for sensing radioactivity inside Sellafield are so sensitive
that alarms are sometimes triggered by the materials
from which the plant is built. Where the materials are
local and the construction reasonably recent, this may
not be proof of *great* sensitivity, after all.) Perhaps the
residues, filtering through the sand over all those years,
have begun to stratify. Plutonium 241 goes critical in
very small quantities. Criticality is a vast release of
energy, deadly to anyone exposed to it. Sellafield as an
environment is unique in history, and I have read
nothing to guide me in imagining what would happen
if, fifteen feet under the sand, a few tablespoons of
volatile isotope converged. It might be an incident that
would spoil a picnic. I have read that at Hanford in
Washington State shifting of soil was required because
wastes had seeped into it and were accumulating dan-
gerously. The analogy here seems potentially useful.
According to the *New Scientist,* in 1986 the Central Policy
Planning Unit of the Ministry of the Environment
suggested that "it would be prudent to place restrictions
on any development along and off the coast near
Sellafield which could disturb the concentrations of
radioactivity building up in mud and silts."[*]

---

[*] "Ministers reveal shortlist for nuclear dumps," *New Scientist,* February 27,
1986, p. 13.

Maybe the beaches at Sellafield had begun to glow in the dark. Islands in the Pacific that were used for atomic testing glowed for years, and contamination levels at Sellafield are like those at testing sites. That would be an excellent reason for hauling the sand away. The matter arose conveniently in the winter months, avoiding any great disruption of the tourist season, at least for American tourists.

Then, too, cancers were accumulating at a rate that would be difficult to ignore. Just at this time a report by James Cutler prepared for Yorkshire Television stimulated a government-appointed commission to look into the leukemia deaths of children in Seascale, the village nearest the plant; the inquiry was headed by Dr. Douglas Black. The Black Report was a response to anecdotal information collected by television journalists, in the event, but the kind of story liable to gain currency even when doctors are legally prohibited from giving out information that is not officially authorized, as in Britain. The physical environs of the plant would have constituted a natural history of contamination, a geolog ical record of a sort, if models and extrapolations were made to take into account tides and seepage and the rest. We must curse the luck that has apparently caused this resource to be lost to science. This is all the more true because the Black Report made a considerable point of stressing the difficulties and uncertainties of establishing dose levels and exposures, and all the more true because a standard defense of practices at Sellafield has been that contaminants disperse in the sea or become somehow fixed to the sea floor.

If Sellafield occupied a cultural terrain where there were such things as liability and culpability, what has happened would appear very like a destruction of evidence. It seems not to have been an act of prudence as that word is normally understood, because nothing was done to decontaminate local houses and shops. I conclude this from the fact that those tested at the time of the Black Inquiry, months later, were found to be contaminated with plutonium and other substances, and this was treated as a surprise. A project was then launched to vacuum houses in the Sellafield area with specially fitted machines, and to do the same in areas well away from nuclear facilities, to determine whether there was any correlation between cancer rates and plutonium in the domestic environment. The project does not seem terribly well designed. The point is to dissociate cancer from radioactive contamination, and the scientific hoovering is to demonstrate, presumably, that there is more radioactive material in houses near nuclear plants than in those at some distance from them. I am at a loss to know what in such information could either surprise or reassure. This demonstration will be made, however, and cancer rates compared, to illustrate that cancer can flourish unabetted by a nuclear power plant. Again, while this can no doubt be proved, there is nothing here to surprise or reassure, either. It is known that near Sellafield the rate of childhood leukemia exceeds the national average by ten times. Where comparable anomalies occur seemingly without exposure to radiation implicated in the deaths of the Cumbrian children, it should be incumbent upon health authorities to look for the causes of these other anomalies. They may be the quirks of statistics, or they may be viruses, or they may be proximity to a chemical waste disposal plant, or to a hospital, since these have been found to dump radioactive iodine used for diagnostic testing into the water system, or to any of the industries that pour detritus into the air and into rivers and estuaries, or even proximity to a rail line along which wastes would pass on the way to Sellafield, or any combination of these factors. If the health consequences of Sellafield blend into larger patterns of public health in Britain, it is because the environmental practices of the nuclear industry are consistent with those of other industries in that befouled country. Finally, the Black Commission concluded that it could offer a limited assurance of the safety of the Cumbrian environment, and so the issue was more or less resolved, at least to the satisfaction of the government.

Yet sometimes the health consequences of radioactive contamination are explicitly conceded. In the recent case of the Black Report, uranium released from the plant was belatedly acknowledged to be a lethal carcinogen through the following sequence of events. Dr. Black, inquiring into the deaths from leukemia among children in the village of Seascale, concluded that emissions from the Sellafield plant could not be blamed because the number of deaths was in *excess* of the number he felt could be projected from the emissions that were supposed to have occurred. According to him, emissions would have had to be forty times as great to account for the actual rate of death. Now, this line of reasoning was ingenious rather than persuasive, in the view of many. The Black Report, with its measured (that is, equivocal) reassurance, seemed open to doubt.

Dr. Black, in a reply to critics, wrote: "Since the report came out, we have been notified of further cases, and indeed asked to include them. We cannot validly do this until the figures in other electoral wards have been

brought up to date for comparison."* Since so much is
made by Sellafield's defenders of the fact that the
number of child deaths—ten—is too small to be significant, though in so small a population it yields an
excess rate of 1,000 percent the national average, it is
strange to minimize the significance of new cases, appearing within months of the publication of Dr. Black's
report. If other electoral wards are in so unhappy a state
as to cushion this rate of excess, God help them.

Dr. Black explains that a cause-and-effect relation
between radiation and leukemia will be established if
lower emissions, which he says are being achieved, bring
lower rates of illness. In other words, future leukemia
excesses will exonerate the plant, as present ones have
done. This is such a pretty piece of reasoning, I will not
spoil it with talk of half-lives.

But just in the nick of time a man came forward with
information which saved the day. In the fifties, when the
plant was still under the management of the United
Kingdom Atomic Energy Authority, a release of uranium occurred which was, uncannily, forty times as
great as had been shown in the records.† The man who
came forward was a former employee, who had left
years before, disgruntled by the fact that releases of
radioactivity from the plant were higher than acknowledged. He and a colleague tested the levels of radioactivity in their homes, found them unsettlingly high, and
decided to leave the area. But he surfaced to make his
telling revelation about an incident in which uranium
was released into the environment. The information

---

* "Sellafield: The nuclear legacy," Dr. Douglas Black, *New Scientist*, March 7,
1985, pp. 12–13.
† "New Sellafield scandal: Government admits true level of radiation was
concealed for 30 years," *The Sunday Times*, February 16, 1986, p. 1.

was opportune in a number of ways. It tended to confirm the accuracy of the Black Inquiry's projections. It associated leukemia with radioactive contamination, but it located the source of the anomaly in a single, discrete episode of contamination. Out of good nature I do not dwell on the persistence of uranium in the environment.

By implication this one episode of contamination being exactly sufficient to satisfy the Black Inquiry's projections, there is nothing else the inquiry failed to take into account. While the plant is implicated in these deaths of children, the rest of the information it gave the inquiry about its operations was at the same time vindicated. (It is sometimes reported that the NRPB, that body whose frequent service to the public has by now made it familiar to my readers, supplied the inquiry with its figures. But if they missed this decisive infusion of uranium into the environment, they must be substantially dependent on industry figures in any case.) Thrift may well have been a factor in the design of the investigation. And in fairness, it had only recently seemed prudent to the government to decontaminate the local beaches, as I have said, and this would necessarily have reduced the value of the area around Sellafield as a source of information, though not to the point where industry estimates need have become the exclusive source.

One feels continuously a sense of lost opportunity. For example, if it seemed appropriate to these inquirers to reason from an excess of childhood leukemia, over and above what their figures led them to predict, to an exoneration of the plant as the cause of leukemia, and if the discovery of the release of uranium undercut this argument by appearing to account precisely for these deaths, could not that first happy conclusion, that the plant was not to blame, have been rescued by drawing attention to the fact that there are elevated rates of leukemia in other villages around Sellafield, and up and down the coast? If excess is exculpatory, then Sellafield is clearly as benign as a clover patch.

As it is, the question has been left in obscurity. Why should a release of uranium that occurred in the fifties have had this dreadful impact on children whose parents were children at the time? If it suggests either chromosome damage or the retention of radioactive substances in the bodies of young women which affect fetal development, then the contamination should manifest itself in other forms besides leukemia. The uranium was apparently vented into the air. Therefore lung cancer would be a likely aftereffect, and the delay in its onset comprehensible. However, only one group of leukemia deaths in one village were within the limits of the study, so other forms of impact of radioactivity were neither sought out nor taken into account where they made themselves manifest. I lay myself open to the charge of cynicism by suggesting that this particular emission was granted its special importance because it occurred under the old management, before BNF took control of the plant. The imputation of carelessness, of bad record keeping, is cast back on the UKAEA, and the present management is unsmirched.

Oddly enough, only days after Dr. Black's results had been, in essential ways, shored up by the discovery of an emission of uranium sufficient to account, by his system of reckoning, for the leukemia deaths of the children of Seascale—a release of uranium from the plant occurred twenty-two times greater than that to which these deaths of children had been more or less attributed. How did the management respond? With public assurances—the Irish were making a fuss—that the release had been approved by the government, was wholly intentional, and presented no threat to anyone. After all, according to former BNF chairman Con Allday, uranium is the most common element in the earth's crust.* He informed the public that the Irish Sea is full of many thousands of tons of naturally occurring uranium. Therefore, another half ton of Sellafield uranium could hardly matter. The (unnamed) chemical plant up the coast releases as much every day—a fact never taken into account in calculating radiation doses, so far as I can discover. In conclusion he laid anxiety about Sellafield to "fear born of ignorance." He does not say whose ignorance inspires this fear.

Other aspects of the nuclear issue are as thoroughly nonsensical. It is said that refusal by the Seamen's Union to man dump ships has ended nuclear waste disposal by Britain into the open sea in the last few years. Since international agreements to stop such dumping have been ignored routinely, there is no great reason to imagine that the action of a labor union will have had a restraining influence, especially on Mrs. Thatcher. The advantage to the government of this action is that it creates obscurity around the situation without the government's having to disavow its policy, should it resume dumping or be found never to have stopped.

---

* "Sellafield: Switch on to the positive," Con Allday, *The Times*, February 20, 1986, p. 12.

In any case, the merits and demerits of ocean dumping from ships—the kind that has supposedly been desisted from—are mulled over in the press as gravely as anyone could wish, though not altogether usefully. The complexities of underground storage are explained with reference to the fact that high-level wastes must be isolated for thousands of years. This information comes as a little shock to one aware of disposal practices at Sellafield, as do the qualms about dumping in the open sea. Jim Slater, former head of the Seamen's Union, spoke of organizing industrial action against Sellafield, and Miss Jean Emery, a leader of Cumbrians Opposed to a Radioactive Environment, has pointed out the absurdity of fretting over the malign consequences of ocean dumping when the quantity that has been put in the Irish Sea from Sellafield is twice as great as that dumped over the sides of ships.* But in general the press seems content to leave all this unreconciled. Despite the supposed halt to the practice, press reports of "stolen minutes" from a meeting of the ministry whose function is to approve dumping at sea record anxiety that the loading of an oversized container, one of special thickness, would tip off the press that plutonium was being dumped, and that this would in turn shake public confidence in ocean dumping.† This is very odd, this glimpse of a government bundling plutonium up in an especially heavy containment and then still chary of being seen to put it in the sea at all. This same ministry has approved all the uncontained disposal that occurs from Sellafield. A cynic might wonder again if this image of a cautious and stalemated government has

---

* Letter, *The Guardian*, September 21, 1983, p. 10.

† "Angry Whitehall stays silent on nuclear waste dumping plans," Paul Brown, *The Guardian*, September 2, 1983, p. 22.

been planted to create characterizing detail at odds with the plain, brute persistence of actual policy.

While British scientists study the relative merits of bores in shale or granite, salt mines, vitrification, or implantation in the seabed as disposal methods suited to materials which must be isolated for periods significant even on geologic time scales, other British scientists ponder the fact that the human placenta has not proved a sufficient protection for the human fetus from plutonium ingested by the mother. Granite is inappropriate to contain plutonium because water can pass through it. The inappropriateness of the placenta for the same function apparently eludes scientific understanding. Unfortunately, while the deficiencies of granite, and doubts about other methods of long-term isolation, have delayed the development of these methods, the same prudential concern has not prevented the disposal practices which rely altogether on frail human flesh. What, after all, should be protected from a notorious mutagen if not a human fetus? This is clearly another instance where industrial practice has run ahead of scientific knowledge, if not in fact away from it. It may be germane here to point out again the great economic advantages entailed in flushing plutonium into the environment. If thrift is a factor, any other method will be hard pressed to compete, more especially now that the horse is out of the barn.

The British ponder costly strategies for disposing of nuclear waste, nuclear power being the only viable long-term energy source for a country that is closing down its coalfields and selling its oil abroad. Faint hearts are scolded for refusing to deal with this hard reality. No mention is made, of course, of the fact that Britain goes looking for trouble, first by soliciting foreign custom for their disposal industry, second by using reprocessing as a disposal method, when the solvents involved multiply the volume of toxic waste more than a hundred times, and third by failing to invest in new plants, which at one time could have set some bounds to the dreadfulness of the enterprise by limiting leaks and spills. The nuclear waste disposal industry, also known as the plutonium industry, slipstreams behind nuclear power as the price that must be paid for industrial vigor. No one seems to dwell upon the fact that the price is paid in Britain for industrial vigor in Germany and Japan, Sweden, Switzerland, and Italy.

Of course they do not bear this cost alone. World commerce in toxins, like every kind of commerce, must suffer from accident and spillage. A traffic in waste destined to end up in the sea is not likely to be obsessively cautious. And then it does end up in the sea, just off the coast of Europe. While Europeans make protesting noises from time to time, their governments pay for these services. The egregiousness of Britain's industrial offenses simply reflects the international role that has been delegated to it, that its peculiar notions of self-interest have caused it to seize upon. Meanwhile, Sweden, a Sellafield client, is constructing a state-of-the-art subsea depository that may be in fact more immune to accidents than most human contrivances. It sounds very impressive, and if it should fail, the wastes so cautiously isolated will at worst only mingle with the Swedish wastes that pass through the pipeline at Sellafield.

I do not know whether I am describing the kind of dissociated behavior that would come with genuine denial, or simply a public-relations stunt, which plays shrewdly on a sad tendency in the public to cling to any little sign of competence on the part of those entrusted with their well-being. On the face of it, all the shielding and tunneling and vitrifying are predicated upon calculations of the dangers of these substances which take them to be extraordinarily great and persistent. So the experience the Europeans have had living alongside seas contaminated with the entire range of radioactive substances produced in reactor cores has not led their specialists to take a more sanguine view of their impact on the environment. This seems to me a fact worthy of note, in light of continuous British assurances that no harm has been done.

Clearly major questions have never been resolved concerning the rights of a national government toward the people and the terrain entrusted to its care. To dispose of either, to sell the health and posterity of one, the habitability of the other, for money, is a perfection of high-handedness beside which all other examples pale. Even to the extent that the mass of people can be thought of as entering into this bargain freely and knowingly, they have sold—for employment, or for some notion of national interest—the well-being of their descendants, which was never theirs to sell, and in the short or medium term, the well-being of the descendants of every mote of life that stirs on the face of the earth. If this has happened in a society which can be called, in any degree, open, free, and democratic, then we had better look at it very seriously indeed. Our own open, free, and democratic country lives in an informational vacuum that makes us a danger to ourselves and a terror to everyone else. No one is any freer than he wishes to be. The apparatus of democracy becomes a sort of Soviet constitution in every instance where there is no will to animate it.

The British are amazingly docile. It is a trait they admire in themselves, and for which they are admired. They have been set apart, among all the developed nations, to endure the insupportable, and they have done it with the quietness and goodwill for which they are legendary. We have justified our reputation for impenetrable ignorance, meanwhile, winging in to drop a tear on the grave of Dorothy Wordsworth and snap a few photos of a gentler world. For forty years, since the end of the Second World War, people have asked how such vile things could have happened as those that deviled Europe in the thirties and forties. The answer is, because anything can happen.

American books on nuclear issues usually omit to mention Britain at all. Jonathan Schell's *The Fate of the Earth* is a distinguished recent example of this tendency. This earnest call to repentance sees nuclear war between the Soviet Union and the United States as the one great peril to the world's survival—implying one great solution, that we "put aside our fainthearted excuses, and rise up to cleanse the earth of nuclear weapons."[*] It is as if history proceeded by referendum, and a grand exertion of collective goodness would put the planet out of harm's way. I have the greatest respect for Schell's religious and democratic zeal, but there is a tendency among committed democrats like us to believe all significant problems must be somehow suited to our solutions, as our pious elders thought their trials were always suited to their strengths. Cleansing the world of

[*] P. 231.

weapons is a relatively simple problem beside cleansing the sea of tons of radioactive sludge, and cleansing the air and the earth, and discovering and limiting the varieties of harm already done. Putting wastes into the sea has been the work of a few bureaucrats. Taking them back out again will be impossible, no matter how aroused and enlightened public opinion might someday become. The problem has been and is now outside democratic political control, first of all because books about nuclear issues do not tell the public the problem exists.

It is a very comfortable thing to think that the greatest threat to the world is a decision still to be made, which may never be made—that is, the decision to engage in nuclear warfare. Sadly, the truth is quite otherwise. The earth has been under nuclear attack for almost half a century.

Mr. Schell explains that, if a nuclear weapon destroyed a nuclear power plant, the radioactive material from the core and the wastes of the plant would be much more virulent and persistent than fallout from the bomb that destroyed it. Then imagine ripe old cores broken down chemically and poured into the environment through a pipeline, or through chimneys and smokestacks. This happens routinely, along the coasts of England and Scotland, and along the coast of France.

Clearly it is not meaningful to say that any sort of permission giving on the part of the public, such as is implied in the existence of nuclear weapons, according to Mr. Schell, lies behind this waste dumping. The people exposed to it are assured that it is not especially harmful. Books and movements which define nuclear peril primarily and even exclusively in terms of nuclear weapons and superpower rivalry confirm these assurances. British people have no grounds whatever to imagine that their situation, notorious as it is, would not impinge on the awareness of a writer who had undertaken so great a subject as the fate of the earth. They must assume therefore that if their radioactive sea does not merit a mention, it cannot be so great a problem after all. I do not believe that Mr. Schell has intentionally excluded information that would complicate the grand simplicity of his thesis. I think he is among those legions who are emotionally incapable of accepting the historical importance of stupidity and furtiveness.

Mr. Schell locates our problems in national sovereignty, by which he means a sort of national self-love, so potent as to make us contemplate a defense that would destroy us. I locate them in the kind of sovereignty that has always been expressed in exploiting and disposing of the lives over which history and accident have given "governments" authority. The fact is that the world public arrives at this parlous moment with a grinding history behind it, badly educated, starved of information, full of sad old fears and desperate loyalties, injured in its self-regard, acculturated to docility and stoicism. The world's most favored public, our own, is educated thoroughly and badly, starved of information, and flattered as to its own importance, while it is made incompetent in the use of the power it has. There is no agora, where issues are really sorted out on their merits and decisions are made which, at best and worst, give permission to political leaders to carry out policies the public has approved. This model assumes information of a quality that is by no means readily available to us. It assumes a reasonableness and objectivity which allow information to be taken in and assimilated to our understanding, and in this we are also thoroughly deficient.

If the world were as Mr. Schell represents it, a place where we make our problems and can unmake them, a place where all those warheads represent public hostility toward the Soviet Union, and a new gospel of love can therefore free us of them, the world would be very simple, simpler than any city, or family, or psyche, or dream. The hostility of Americans toward Russians is an invention of polemicists. If the Soviet Union is author-itarian, so are most countries. While atheism is espoused by its government, religion seems to flourish among its people. Western European cultures, by contrast, are atheist in fact, at street level, and that has never struck us as any abomination or unbridgeable divide. Like most things, it has never struck us at all. If Russia ceased to appear to us as a threat, we would probably simply forget it, as we do most of the world most of the time. The tendency of this country to be engrossed in itself makes it ill suited to sustaining large-scale, long-term interest of any kind in the outside world. But we are told constantly that the government of the Soviet Union has aggressive intentions, and we remember just enough modern history to know what that can mean. Presum-ably the Russian state of mind is some version of this, *mutatis mutandis*, and people may well unite to eliminate nuclear weapons, at least in the countries that acknowl-edge having them and, unlike Britain and France, are willing to submit to international agreements to control them.

Nuclear weapons can be produced at short notice by anyone in possession of fissionable materials, of course, but even if they are not simply replaced in secret after they are destroyed in public, fissionable materials will continue to be produced, and toxic and radioactive materials of even greater virulence than those used in bombs, through the routine functioning of nuclear power plants, so many of which were built to produce bomb-grade plutonium as well as electricity, and will continue to produce it for as long as they are used for power generation. So at best these diabolical substances will accumulate as wastes rather than as warheads, but more toxic because they will not be dissipated in the upper atmosphere but will burn or leak into the ground or simply be buried or dumped somewhere, as in fact most wastes have been for forty years. In the long term it will not matter whether national sovereignties destroy their "enemies" or merely themselves and their neighbors. The fate of the earth will be the same.

An October 1987 article in *The New York Times*[*] informed those of its readers capable of absorbing the information that an agreement, classified along with the analysis which supported it, had been signed by our Secretaries of State and Energy, to permit flights carrying plutonium from Britain and France to Japan to land and refuel in Anchorage. This is the kind of situation in which one regrets that there is not more attachment to "national sovereignty," in Jonathan Schell's sense of the phrase. The governor of Alaska has sued to have the shipments blocked, and has failed to win a restraining order. How unfortunate for him that the issue arose just when other stories of greater urgency, for example the television evangelism scandals, were filling the front pages of America's newspapers. The governor's suit charges that "thousands of pounds" of plutonium will pass through Anchorage, and quotes a physicist from the University of Michigan to the effect that "plutonium

---

[*] "Alaska Seeks Halt to Plutonium Plan," Hal Spencer, *The New York Times*, October 4, 1987, p. 33.

is one of the most, 'if not the most,' toxic substances
known to humans. Inhaling a microscopic speck could
lead to cancer." That is, of course, the usual formula for
describing the toxicity of plutonium.

I doubt that our Secretaries of State and Energy have
considered and signed such an agreement casually. It is
entirely possible that they signed it to prevent the
refuelings from occurring in Seattle or Los Angeles,
without approval, and without special security measures.
After all, this commerce is being run by people who see
no harm in "taking a bit of ploot." Small amounts of
plutonium would be easy to conceal, in the absence of
any special precautions. A letter to the governor, Steve
Cowper, signed by Secretary of State George P. Shultz,
said that approval for these flights "will be conditioned
upon a number of safety requirements such as transfer
exclusively by air (to minimize time spent in international transit), use of a cask certified to withstand a
crash, armed guards, redundant communications and
detailed contingency plans." If the conditions the Secretary sets out are not met, how will he know? If they are
met in 10 percent of shipments, while the other 90 are
stowed away in other passenger or cargo flights, will he
be the wiser? Most European power plants are built on
national boundaries. Therefore any accident will be half
the problem of another government. Aside from its
being an interesting comment on their view of the safety
of their own industries and a telling comment on all the
gasps of surprise, at the time of Chernobyl, that nuclear
reactor accidents know no boundaries, it reveals a
certain willingness to let foreigners bear the brunt of
risky policies. If plutonium burned in an airliner crash,
would anyone know? Would the discovery of these
residues afterward be laid to a non-domestic source? I

suspect the real nature of this "agreement" is simply a plea to the Europeans and Japanese to tell us what they are doing and when they are doing it. The threat to end the permission only threatens us with uncontrolled movement of plutonium through our hemisphere, no problem to the Japanese, who are accustomed to seeing their wastes dumped into European coastal waters, and no problem to the Europeans, who consider this an excellent business to be in despite self-inflicted contamination on a scale no accident could visit on us here.

A second *New York Times* article, published a few weeks later, described a report sent to the Congress by the Defense Department, warning that increased production and use of plutonium, and increased international shipments of radioactive materials, will increase the risks of theft, diversion, and terrorist acts.[*] The article explains quaintly, "The United States produces plutonium only at military installations for use in weapons. France and other countries, however, are exploring the feasibility of breeder reactors to produce plutonium commercially to fuel other reactors or for weapons." In other words, commercial production of plutonium has not yet begun, or so anyone would infer who did not know better.

The article informs us that "International agreements and American law govern the security provisions enforced when plutonium is moved." Well, this is something one would never learn from reading the British press. In nothing is a more sublime autonomy displayed than in the United Kingdom's dealings in plutonium. The bomb plant at Sellafield was created in the first instance in defiance of American attempts to control

---

[*] "Rising Nuclear Trade Stirs Fear of Terrorism," John H. Cushman, Jr., *The New York Times*, November 5, 1987, p. 5.

nuclear proliferation, and nothing that has happened subsequently indicates any second thoughts. Either international standards mean nothing at all or they mean it is acceptable to ship nuclear wastes across the world to be dumped into British and European waters—which is to say, they mean nothing at all. Their single function seems to be to baffle the Yankees, and that they do very well.

It is worth noting how plutonium and radioactive materials are weapons intrinsically, as the London *Times* editorialist understood in 1976. We cannot close our borders against plutonium because it *is* plutonium, and liable to punish us brutally if we make the attempt. Our sovereignty is overridden by allies under cover of our own poor journalism. Is this the expression of the will of our people? Are they so eager to expedite this disastrous commerce that they would knowingly accept its risks? Of course not.

Except, perhaps, for that numerous new breed of moralist thrown up by this sad age, which will reprove me for criticizing Britain—unheard-of cheek. But we are talking about the world, after all, which history has placed in our most unworthy hands.

The final, visceral loyalty of American "intellectuals" to Europe is racism. The refusal to see the dimensions of phenomena like Sellafield, the refusal to call them by the hard names that fit them, is racism. If you think the Third World is hungry now, wait till the sea is dead.

Of course the United States has been smirched by history. But in the larger scheme, the United States is an invention, like Constantinople, which, if life could be imagined going on, would drift and evolve into other shapes and things in the way of species, clouds, and continents. If I could dream that the world would live so long that our books were lost and our name forgotten, I could feel we had been a good and successful civilization, after all. We give countries kinds of reality they do not have. They do not define the natures or the obligations of the human beings who live in them. Our country allows and encourages us to know nothing. But if we are ignorant, the fault is ours. Increasingly it encourages us, through its educational institutions, press, and popular culture, to consider ourselves knaves and fools. But if we act like fools, the fault is ours.

The recent decline in national self-esteem has led many Americans to invest their emotions offshore, in what they take to be a favorable climate, among solvent institutions. In imagination they have escaped ruin, growing rich as their neighbors grew poor. These people do not want to hear bad news.

But there is a real world, that is really dying, and we had better think about that. My greatest hope, which is a very slender one, is that we will at last find the courage to make ourselves rational and morally autonomous adults, secure enough in the faith that life is good and to be preserved, to recognize the grosser forms of evil and name them and confront them. Who will do it for us? E. P. Thompson? Greenpeace? The Duke of Edinburgh? *The Washington Post?* We have to walk away from this road show, consult with our souls, and find the courage, in ourselves, to see, and perceive, and hear, and understand.

Selected Bibliography

"£700m nuclear deal with Japan near signing," Pearce Wright, *The* (London) *Times*, October 23, 1976, p. 4.

"Fear that nuclear power plans could threaten freedom," Pearce Wright, *The* (London) *Times*, October 28, 1976, p. 4.

"Fears about effect of nuclear power plans," *The* (London) *Times*, October 28, 1976, p. 14.

"The Plutonium Problems," editorial, *The* (London) *Times*, October 28, 1976, p. 17.

"Mr. Benn rules out need for quick decision over fast nuclear reactor programme," *The* (London) *Times*, December 14, 1976, p. 4.

"France bans sale of atom fuel plants," *The* (London) *Times*, December 17, 1976, p. 5.

"Government survey of nuclear waste needs," Pearce Wright, *The* (London) *Times*, December 17, 1976, p. 2.

"Windscale to check deaths records," Michael Morris, *The Guardian*, July 14, 1977, p. 5.

"Nuclear Fuels answers storage query," Malcolm Pithers, *The Guardian*, July 15, 1977, p. 3.

"Tip of A-waste iceberg," Richard Norton-Taylor, *The Guardian*, July 26, 1977, p. 2.

"Terrorist peril in nuclear waste," *The Guardian*, July 27, 1977, p. 3.

"Another tough customer for British reactors," *New Statesman*, March 18, 1983, p. 5.

"A new kind of nuclear victim," Rob Edwards, *New Statesman*, July 22, 1983, p. 8.

"Wasting the ocean," Rob Edwards, *New Statesman*, July 22, 1983, p. 6.

"Angry Whitehall stays silent on nuclear waste dumping plans," Paul Brown, *The Guardian*, September 2, 1983, p. 22.

" 'Awkward questions' about nuclear waste dumping," Rob Edwards, *New Statesman*, September 2, 1983, p. 4.

"Coroner halts weapons scientist's cremation," *The Guardian*, September 2, 1983, p. 3.

"The monster that won't lie down," Duncan Campbell, *New Statesman*, September 2, 1983, pp. 8–9.

"£5m research on ways to curb pollution by acid rain: Study could give ammunition to nuclear power lobby," Andrew Moncur, *The Guardian*, p. 2.

"Slater fears nuclear waste may be dumped on sea bed in submarine," *The Guardian*, September 10, 1983, p. 2.

"Nuclear waste sea dumping plans revived," Paul Brown and Richard Norton-Taylor, *The Guardian*, September 14, 1983, p. 1.

"New law on dumping may be needed," Paul Brown, *The Guardian*, September 15, 1983, p. 28.

"Peril in the deep," editorial, *The Guardian*, September 15, 1983, p. 28.

Letter, Jean Emery, *The Guardian*, September 21, 1983, p. 10.

"Leukemia study finds cluster of cases near Sizewell nuclear power station," Roger Milne, *The Guardian*, September 28, 1983, p. 2.

"New waste dumps," Richard Norton-Taylor, *The Guardian*, September 28, 1983, p. 3.

"New Windscale report hints at 33 deaths," Anthony Tucker, *The Guardian*, September 28, 1983, p. 3.

"Nuclear waste sent back after union ban," Penny Chorlton, *The Guardian*, September 28, 1983, p. 3.

"The dirtiest nation on Earth," James Erlichman, *The Guardian*, October 3, 1983, p. 14.

"Ministry denies plans to dump plutonium at sea," Richard Norton-Taylor, *The Guardian*, October 3, 1983, p. 14.

"The first chair on the sea bed," editorial, *The* (London) *Times*, October 6, 1983, p. 17.

"Nuclear waste discharges into sea to be reduced," David Fairhall and Frank Scimone, *The Guardian*, October 6, 1983, p. 2.

"Nuclear firm told to stop discharge into sea," Frank Scimone, *The Guardian*, October 8, 1983, p. 2.

"Nuclear industry policy," letter, Professor Ian Fells, *The* (London) *Times*, October 11, 1983, p. 13.

"Sellafield cuts plutonium discharges," *New Scientist*, October 13, 1983, p. 73.

"Sizewell leukemia inquiry," Anthony Tucker, *The Guardian*, October 21, 1983, p. 2.

"Nuclear accident victim's 19 year agony after the 'big blow-out,' " Ted Harrison and Geoffrey Lean, *The Observer*, October 23, 1983, p. 3.

"Nuclear dumps safety guidelines ignore 'cocktail' factor," Anthony Tucker, *The Guardian*, October 27, 1983, p. 2.

"Sea dumping plan intact," John Gapper, *The Guardian*, October 28, 1983, p. 8.

"Children near Windscale have high cancer levels," Geoffrey Lean, *The Observer*, October 30, 1983, p. 1.

"Windscale leukemia link denied: Plutonium found in house dust brings fear on radiation levels," Anthony Tucker, *The Guardian*, October 31, 1983, p. 24.

"BNF attacks 'one-sided cancer' film publicity," David Pallister, *The Guardian*, November 1, 1983, p. 3.

"Inquiry starts into cases of leukemia near nuclear power stations," Pearce Wright, *The* (London) *Times*, November 1, 1983, p. 2.

"Fuelling nuclear fears," editorial, *The* (London) *Times*, November 2, 1983, p. 15.

"Windscale's dirty linen," television review, Nancy Banks-Smith, *The Guardian*, November 2, 1983, p. 9.

"Inquiry into Windscale cancer rate," Pearce Wright, *The* (London) *Times*, November 3, 1983, p. 2.

"Nuclear cancer link for inquiry," Julian Langdon, *The Guardian*, November 3, 1983, p. 1.

"Radiation fears soothed," *The Guardian*, November 3, 1983, p. 23.

"Dangers and defenses in Sellafield plant's emissions," then-BNF chairman Con Allday, *The* (London) *Times*, November 4, 1983, p. 13.

"Windscale dilemma: Is radiation ever safe?" Robin McKie, *The Observer*, November 6, 1983, p. 2.

"Nuclear shipper and wife found dead," Michael Morris, *The Guardian*, November 9, 1983, p. 3.

"Radioactive waste put on council tip," Mark Rosselli, *The* (London) *Times*, November 11, 1983, p. 2.

"Unions may link to halt transport of nuclear waste," Robin McKie, *The Observer*, November 13, 1983, p. 2.

"Seascale warning led to sack for scientist," *The Guardian*, November 14, 1983, p. 1.

"Windscale panel to investigate a hundred deaths," *The* (London) *Times*, November 14, 1983, p. 1.

"Nuclear protest charges," *The Guardian*, November 15, 1983, p. 3.

"Divers near Windscale pipe 'contaminated,' " Paul Brown, *The Guardian*, November 17, 1983, p. 2.

"Leukemia 'cannot be blamed on Windscale,' " *The Guardian*, November 17, 1983, p. 2.

"Windscale: British Nuclear Foul-up Limited," James Cutler, *New Statesman*, November 18, 1983, p. 8.

"Windscale crackdown on way," Geoffrey Lean, *The Observer*, November 20, 1983, p. 1.

"Government 'Sellafield fears,' " David Nicholson-Lord, *The* (London) *Times*, November 21, 1983, p. 1.

"Windscale leak 'is bigger than admitted,' " David Pallister, *The Guardian*, November 21, 1983, p. 1.

Untitled essay, Tony Benn, *The Guardian*, November 21, 1983, p. 15.

"Inquiry into radioactive leak," *The* (London) *Times*, November 22, 1983, p. 4.

"N-quiz for minister," *The Guardian*, November 22, 1983, p. 25.

"Sellafield may face outside inquiry into spill," Colin Brown, *The Guardian*, November 22, 1983, p. 1.

"Clouds over Sellafield," editorial, *The Guardian*, November 23, 1983, p. 12.

"Nuclear waste rebels defy ban," Paul Brown and Michael Morris, *The Guardian*, November 23, 1983, p. 1.

"Indecent exposure," Anthony Tucker, *The Guardian*, November 24, 1983, p. 23.

"Call for atom plant probe," Geoffrey Lean, *The Observer*, December 4, 1983, p. 5.

"Sellafield staff angry," *The* (London) *Times*, December 9, 1983, p. 2.

"Windscale: Six years later . . ." Rob Edwards, *New Statesman*, December 9, 1983, p. 11.

"Jenkin defends discharge into sea at Sellafield," Ronald Faux, *The* (London) *Times*, December 10, 1983, p. 3.

"Nuclear watchdog condemns Sellafield safety," Richard Norton-Taylor and Peter Hetherington, *The Guardian*, December 10, 1983, p. 1.

"Windscale beach pollution blunder," Geoffrey Lean, *The Observer*, December 11, 1983, p. 2.

"Sellafield faces prosecution over leaks," *New Scientist*, December 15, 1983, p. 791.

"Windscale's pipeline," Peter Danckwerts, *New Scientist*, December 15, 1983, p. 883.

"Doctors say 'spies' are leaking hospital secrets," *The Guardian*, December 16, 1983, p. 2.

"Leukemia questioning vetoed by inspector," Roger Milne, *The Guardian*, December 16, 1983, p. 2.

"Mistake led to nuclear waste in sea," Michael Morris, *The Guardian*, December 16, 1983, p. 1.

"Valve error caused Sellafield leak," Ronald Faux, *The* (London) *Times*, December 16, 1983, p. 3.

"Radioactive leak case referred to the DPP," *The* (London) *Times*, December 21, 1983, p. 4.

"Radioactive leak at Sellafield may lead to prosecution," Philip Webster, *The* (London) *Times,* December 22, 1983, p. 1.

"Radioactive leak reported to the DPP, Jenkin tells Commons," *The Guardian,* December 22, 1983, p. 14.

"Sellafield leak is referred to DPP: Scientists worried by 'strange' pattern of sea contamination," Anthony Tucker, *The Guardian,* December 22, 1983, p. 1.

"BNFL pays £21,000 in leukemia case," Michael Morris, *The Guardian,* December 23, 1983, p. 2.

"Plan to sink waste in seabed," Paul Keel, *The Guardian,* December 23, 1983, p. 1.

"A-plants: A hardcore problem in the making," Barbara von Ow, *The* (London) *Times,* December 27, 1983, p. 8.

"Sellafield cancer victims given compensation but liability not admitted," *The* (London) *Times,* January 4, 1984, p. 3.

"BNFL denies 'cover-up' claim over Sellafield compensation payments," Paul Keel, *The Guardian,* January 5, 1984, p. 2.

"Sellafield claim by MP is rejected," Ronald Faux, *The* (London) *Times,* January 5, 1984, p. 2.

"Whitehall withholds nuclear papers," Richard Norton-Taylor, *The Guardian,* January 10, 1984, p. 4.

"Greenpeace may defy Sellafield pipe ban," John Witherow, *The* (London) *Times,* January 14, 1984, p. 3.

"Partial victory for Greenpeace despite ruling," Penny Chorlton, *The Guardian,* January 14, 1984, p. 2.

"Sellafield beach risk still very high," Paul Brown, *The Guardian,* January 15, 1984, p. 28.

"Cancer inquiry leader will not visit village," Paul Brown, *The Guardian,* January 20, 1984, p. 2.

"Radiation leak at top-secret plant," Thomson Prentice, *The* (London) *Times,* January 20, 1984, p. 1.

"Cancer worry at fourth atom plant," Geoffrey Lean, *The Observer,* January 22, 1984, p. 7.

"Argument over 'safe' radiation," Anthony Tucker, *The Guardian,* January 26, 1984, p. 26.

"Deformed babies blamed on Windscale fire," Roger Milne, *New Scientist,* January 26, 1984, p. 7.

"Warning of risk to Sellafield children," Andrew Veitch, *The Guardian,* January 27, 1984, p. 4.

"Europe's N-waste may go to Billingham," Malcolm Wright, *The Guardian,* January 30, 1984, p. 2.

"Radiation check demanded," Paul Keel and Michael Morris, *The Guardian,* February 1, 1984, p. 2.

"Windscale: Backing for leukemia charge," *New Scientist,* February 2, 1984, p. 4.

"Sellafield waste traces 'found under Arctic ice,' " David Fairhall, *The Guardian,* February 3, 1984, p. 4.

"Cancer patients 'refused places on waiting lists,' " Penny Chorlton, *The Guardian,* February 6, 1984, p. 2.

"Ecology groups raided by police over leak," Richard Norton-Taylor, *The Guardian,* February 7, 1984, p. 30.

"Profit motive fears over nuclear waste research," Malcolm Wright, *The Guardian,* February 10, 1984, p. 2.

"Radioactivity: Polluting suburbs," Andrew Tyler, *New Statesman,* February 10, 1984, pp. 10, 12.

"Safety flaws exposed at Windscale," Geoffrey Lean, *The Observer,* February 12, 1984, p. 2.

"Sellafield inquiry reports criticize process of removing radioactive refuse," Ronald Faux, *The* (London) *Times,* February 13, 1984, p. 2.

"A disaster that almost happened?" editorial, *The Guardian,* February 15, 1984, p. 10.

"Errors that let waste into sea," Paul Brown, *The Guardian,* February 15, 1984, p. 2.

"Government blames Sellafield chiefs for leak," Paul Brown and Colin Brown, *The Guardian,* February 15, 1984, p. 1.

"Nuclear waste reports warn public to avoid Sellafield beaches," Pearce Wright, *The* (London) *Times,* February 15, 1984, p. 2.

"Promises leave the locals unmoved," Michael Morris, *The Guardian,* February 15, 1984, p. 2.

"Sellafield controls found wanting," Paul Brown, *The Guardian*, February 15, 1984, p. 2.

"Sellafield catalogue of errors," Pearce Wright, *The* (London) *Times*, February 15, 1984, p. 1.

"Fuel for fear," editorial, *The* (London) *Times*, February 16, 1984, p. 13.

"Union stands by Sellafield staff," Paul Brown, *The Guardian*, February 16, 1984, p. 3.

"Open verdict on cancer victim contaminated by nuclear leaks," *The Guardian*, February 17, 1984, p. 2.

"Dublin demands Sellafield action," Ian Black, *The Guardian*, February 18, 1984, p. 28.

"Warnings go up on nuclear site beaches," Michael Morris, *The Guardian*, February 18, 1984, p. 3.

"Nuclear Fuels sale at risk," Robin McKie and Geoffrey Lean, *The Observer*, February 19, 1984, p. 3.

"CND accuses industry of plutonium trade cover-up," David Fairhall, *The Guardian*, February 20, 1984, p. 3.

"Facts concealed on nuclear fuel supply," Pearce Wright, *The* (London) *Times*, February 20, 1984, p. 2.

"Hopes pinned on the young blood in Sellafield's front line," letter, then-BNF chairman Con Allday, *The Guardian*, February 20, 1984, p. 12.

"The greenhouse effect," editorial, *The* (London) *Times*, February 23, 1984, p. 13.

"Pollution fight bogged down," Paul Keel and Martin Wainwright, *The Guardian*, February 23, 1984, p. 6.

"Plutonium: Extra scrutiny," Duncan Campbell, *New Statesman*, February 24, 1984, pp. 11–12.

"No agreement on nuclear dumping," Pearce Wright, *The* (London) *Times*, February 25, 1984, p. 2.

" 'Chemical dustbin' fears over site near Sellafield," Michael Morris, *The Guardian*, February 27, 1984, p. 3.

"Britain is urged to reduce pollution," Donald Fields, *The Guardian*, March 2, 1984, p. 9.

"Radioactive contamination of Sellafield beaches increasing, survey shows," Pearce Wright, *The* (London) *Times*, March 2, 1984, p. 5.

"Sellafield denies new leak after tar finds," Michael Morris, *The Guardian*, March 2, 1984, p. 1.

"ICI rejects plan for Billingham nuclear dump," James Erlichman, *The Guardian*, March 6, 1984, p. 30.

"Nuclear 'laundry' tally," *The Guardian*, March 6, 1984, p. 9.

"Sizewell waste plan criticized," Pearce Wright, *The* (London) *Times*, March 8, 1984, p. 5.

"New use for radioactive waste," Pearce Wright, *The* (London) *Times*, March 9, 1984, p. 2.

"More pollution found on coast near Sellafield," *The Guardian*, March 10, 1984, p. 3.

"Pest spray plea," Geoffrey Lean, *The Observer*, March 11, 1984, p. 5.

"Beach contamination at Dounreay," David Hearst, *The Guardian*, March 14, 1984, p. 9.

"Inquiry considers Leiston leukemia," *New Scientist*, March 15, 1984, p. 8.

"UK pollution control goes up in smoke," Geoffrey Lean and Tony Heath, *The Observer*, March 18, 1984, p. 4.

"Beach debris to be monitored by company for radioactivity," Michael Morris, *The Guardian*, March 21, 1984, p. 6.

"US confirms use of British plutonium in its weapons," Richard Norton-Taylor, *The Guardian*, March 22, 1984, p. 2.

"Sellafield safety team checks coast," Michael Morris, *The Guardian*, March 23, 1984, p. 4.

"Sellafield radiation study," *The* (London) *Times*, March 27, 1984, p. 3.

"Doubts over how Sizewell waste will be handled," *The* (London) *Times*, March 29, 1984, p. 2.

"Nordic countries press Britain to cut radioactive discharges," Paul Brown, *The Guardian*, March 29, 1984, p. 3.

"Sellafield cancer evidence 'growing,' " Paul Brown, *The Guardian*, March 29, 1984, p. 2.

"Sellafield cancer link denied by BNFL," David Hearst, *The Guardian*, March 30, 1984, p. 2.

"Two mishaps in 80 days at Windscale," Geoffrey Lean and David Siddall, April 1, 1984, p. 2.

"Radioactive waste found on coast at Whitehaven," Michael Morris, *The Guardian*, April 5, 1984, p. 4.

"Sellafield villagers offered body scans," Pearce Wright, *The* (London) *Times*, April 13, 1984, p. 3.

"Jenkin calls Sellafield beach risk negligible," Paul Brown, *The Guardian*, May 6, 1984, p. 2.

"Britain lags in energy saving," Rod Chapman, *The Guardian*, May 22, 1984, p. 22.

"Beachcombers beware," Ruth Balogh, *New Statesman*, May 25, 1984, p. 6.

"Scots cancer cases surprise Sellafield Inquiry scientists," Andrew Veitch, *The Guardian*, May 25, 1984, p. 4.

"Incidence of leukemia in young persons in west of Scotland," letter, M. A. Heasman, I. W. Kemp, Anne Marie MacLaren, P. Trotter, C. R. Gillis, and D. J. Hole, *The Lancet*, May 26, 1984, pp. 1188–9.

"Cancer 'cluster' round nuclear plant," Rob Edwards, *New Statesman*, June 1, 1984, p. 4.

"Report delay as cancer facts mount," Joan Smith, *New Statesman*, June 8, 1984, p. 5.

"Sellafield launches effluent study," Pearce Wright, *The* (London) *Times*, June 8, 1984, p. 2.

"Unions threaten to block deliveries to plant," Barrie Clement, *The* (London) *Times*, June 8, 1984, p. 2.

"Sellafield discharge criticized," Pearce Wright, *The* (London) *Times*, July 5, 1984, p. 5.

"Cumbria: A special report," Ronald Faux, *The* (London) *Times*, July 6, 1984, p. 14.

"Nuclear weapons: Plutonium goes critical," Rob Edwards, *New Statesman*, July 13, 1984, p. 14.

"Apparent clusters of childhood lymphoid malignancies in Northern

England," letter, A. W. Croft, S. Openshaw, and J. Birch, *The Lancet,* July 14, 1984, pp. 96–7.

"Fuel flask survives 100 mph impact," *The* (London) *Times,* July 18, 1984, p. 3.

"Nuclear link to child cancers," *The Guardian,* July 20, 1984, p. 1.

"Sellafield leukemia allegations supported," Pearce Wright, *The* (London) *Times,* July 21, 1984, p. 3.

"Sellafield reassurance," David Hearst, *The Guardian,* July 21, 1984, p. 3.

"The peril that lurks by the sea," James Cutler, *The Guardian,* July 23, 1984, p. 15.

"Cautious view of Sellafield 'cancer link,' " Paul Denis, *The Guardian,* July 24, 1984, p. 4.

"Leukemia deaths higher at Sellafield," Pearce Wright, *The* (London) *Times,* July 24, 1984, p. 2.

"Lingering particles of unease," editorial, *The Guardian,* July 24, 1984, p. 10.

"Sellafield cancer link 'not proven,' " Paul Brown, *The Guardian,* July 24, 1984, p. 1.

"Routine leukemia tests urged for nuclear staff," Roger Milne, *The Guardian,* July 25, 1984, p ?

"Jury baffled by missing Sellafield records," Ruth Balogh, *New Statesman,* July 27, 1984, p. 6.

"Sellafield birds that fail to breed," Michael Morris, *The Guardian,* July 27, 1984, p. 4.

"London to stay on nuclear fuel route, says BR," David Fairhall and Roger Milne, *The Guardian,* July 28, 1984, p. 4.

"Haughey calls for Sellafield to be closed," Paul Brown and Paul Johnson, *The Guardian,* July 30, 1984, p. 1.

"Key facts 'kept from Sellafield inquest jury,' " Paul Brown, *The Guardian,* July 30, 1984, p. 3.

"CND plutonium plea fails on security grounds," *The Guardian,* August 1, 1984, p. 3.

"Sellafield beaches given safety all-clear," Paul Brown, *The Guardian,* August 1, 1984, p. 1.

"BNFL accused of leak offenses," Paul Brown, *The Guardian*, August 2, 1984, p. 1.

" 'No hazard' if terrorists hit nuclear flasks," *The Guardian*, August 2, 1984, p. 2.

"Firms set up high tech link with Japan to share in pollution control boom," Maggie Brown, *The Guardian*, August 3, 1984, p. 3.

"The main lesson from Sellafield," editorial, *The Lancet*, August 4, 1984, pp. 266–7.

"Acid rain report seeks action on dying rivers," Paul Brown, *The Guardian*, August 9, 1984, p. 2.

"Windscale plutonium on Scottish east coast," Rob Edwards, *New Statesman*, August 13, 1984, p. 5.

"Murder victim attacks plan for Sizewell," *The Guardian*, August 18, 1984, p. 3.

"A Geiger counter set to soothe sleepless nights," Paul Brown, *The Guardian*, August 23, 1984, p. 17.

'Minister in swim off Sellafield," August 23, 1984, p. 4.

"Sellafield leak will slow privatisation," David Simpson, *The Guardian*, August 23, 1984, p. 1.

"N-waste build-up claim," Andrew Veitch, *The Guardian*, August 24, 1984, p. 1.

"Tourist industry: Blackpool beached," Kevin Sutcliffe, *New Statesman*, August 24, 1984, p. 10.

"Leukemia and Sellafield," letters, R. Russell-Jones, G. B. Schofield, *The Lancet*, August 25, 1984, pp. 467–8.

"Uranium gas tanks sank with freighter," Paul Brown and David Hearst, *The Guardian*, August 27, 1984, p. 1.

"Dumping of dials 'a mistake,' " *The Guardian*, August 29, 1984, p. 2.

"Seamen seek to stop nuclear shipments," Paul Brown, *The Guardian*, August 29, 1984, p. 26.

"Soviets 'help to arm US missiles,' " Paul Brown, *The Guardian*, August 29, 1984, p. 26.

"UK cuts costs with Soviet nuclear deal," David Fairhall, *The Guardian*, August 31, 1984, p. 2.

"Fitting in with the locals," Ruth Balogh, *New Statesman,* September 14, 1984, pp. 8–9.

"Britain to attend Australian A-test inquiry," *The* (London) *Times,* November 16, 1984, p. 6.

"Inquiry split on end use of plutonium," Pearce Wright, *The* (London) *Times,* November 28, 1984, p. 2.

"Minister seeks TUC aid in easing nuclear waste delay," Pearce Wright, *The* (London) *Times,* February 11, 1985, p. 4.

"Mystery as blockage shuts down Sellafield plant," *The* (London) *Times,* February 15, 1985, p. 5.

"Radioactive mud protest," *The* (London) *Times,* February 27, 1985, p. 4.

"Inquiry into Murrell murder," Stephen Cook, *The Guardian,* March 4, 1985, p. 3.

"Cancer inquiry," *The* (London) *Times,* March 7, 1985, p. 4.

"Jenkin agrees to independent look at proposals for nuclear waste," Paul Brown, *The Guardian,* March 7, 1985, p. 4.

"Ulster to monitor Sellafield spillage," Paul Johnson, *The Guardian,* March 7, 1985, p. 4.

"Nuclear power case 'made by miner's strike,' " David Fairhall, *The Guardian,* March 8, 1985, p. 26

"Nuclear power 'kept the lights on' during strike," David Young, *The* (London) *Times,* March 8, 1985, p. 2.

"N-waste heads for a million tonnes," Paul Brown, *The Guardian,* March 11, 1985, p. 2.

"Dalyell 'has new clue to activist's murder,' " Paul Keel, *The Guardian,* March 16, 1985, p. 2.

"MP to help Murrell inquiry," Rupert Morris, *The* (London) *Times,* March 16, 1985, p. 2.

"Radioactive disposal and decommissioning of plants reconsidered," Pearce Wright, *The* (London) *Times,* March 18, 1985, p. 4.

"Cancer risk at heart of safety argument," Pearce Wright, *The* (London) *Times,* March 19, 1985, p. 4.

"N-waste storage option," Paul Brown, *The Guardian,* March 19, 1985, p. 2.

"A-test story horrifies Wick," Ronald Faux, *The* (London) *Times*, March 20, 1985, p. 4.

"Atom bomb trials in Britain disclosed," Pat Healy, *The* (London) *Times*, March 21, 1985, p. 2.

"Radioactive discharge to sea will be reduced by new Sellafield plant," Pearce Wright, *The* (London) *Times*, March 21, 1985, p. 5.

"£120,000 paid to family of Sellafield worker," David Rose and Richard Norton-Taylor, *The Guardian*, March 27, 1985, p. 3.

"£120,633 award after radiation-link death," Pearce Wright, *The* (London) *Times*, March 27, 1985, p. 2.

"Public 'ill-informed on issues,' " David Nicholson-Lord, *The* (London) *Times*, March 27, 1985, p. 3.

"Yorkshire might have been atom bomb test site," Anthony Bevins, *The* (London) *Times*, March 28, 1985, p. 2.

"How Britain produced the Bomb," Margaret Gowing, *The Guardian*, April 8, 1985, p. 9.

"Terms of nuclear dumping inquiry 'renege on government promise,' " Paul Brown, *The Guardian*, April 10, 1985, p. 6.

"Sellafield parishes to fight review," Michael Morris, *The Guardian*, April 12, 1985, p. 2.

"British setback in sea dumping report," Paul Brown, *The Guardian*, April 13, 1985, p. 3.

"British uranium stocks now far outstrip demand," Paul Brown, *The Guardian*, April 15, 1985, p. 3.

"Nuclear leak 'did not pose danger to city,' " Paul Brown, *The Guardian*, April 19, 1985, p. 23.

"Murrell murder: New theory," Richard Norton-Taylor, *The Guardian*, April 22, 1985, p. 2.

"Overloaded lorry had nuclear cargo," Paul Brown, *The Guardian*, April 25, 1985, p. 2.

"Statistics, St. Petersburg and Sellafield," John Ennis, *New Scientist*, May 2, 1985, pp. 26–7.

" 'Plutonium food' sought for children," Richard Evans, *The* (London) *Times*, May 21, 1985, p. 1.

Marilynne Robinson   2 5 1

"BNF pollution trial starts," Pearce Wright, *The* (London) *Times,*
June 6, 1985, p. 3.

"Sea waste 'radioactive oily slick,' " *The* (London) *Times,* June 21,
1985, p. 5.

"Sellafield nuclear waste acquittals," *The* (London) *Times,* July 10,
1985, p. 2.

" 'Surprise' over level of radiation leak," *The* (London) *Times,* July 11,
1985, p. 4.

" 'No secret' over tests on Sellafield babies," *The* (London) *Times,* July
29, 1985, p. 3.

"Crime and Punishment," Greenpeace *Examiner,* September 1985, p.
7.

"Uranium scare on Irish coast," Fabian Boyle, *Irish News,* January 28,
1986, p. 1.

"Inquiry begins into leukemias in children," Roger Milne, *New
Scientist,* January 30, 1986, p. 24.

"Plutonium makes waves," Roger Milne, *New Scientist,* January 30,
1986, pp. 55–8.

"Strike follows leak at N-plant," Peter Davenport and Pearce Wright,
The (London) *Times,* February 2, 1986, p. 1.

"Serious incidents double at Dounreay plant," *The Guardian,* February 3, 1986, p. 2.

"Sellafield alert after leak of plutonium," Pearce Wright, *The* (London) *Times,* February 6, 1986, p. 1.

"More tests on staff at Sellafield," *The* (London) *Times,* February 7,
1986, p. 7.

"The nuclear mist," editorial, *The* (London) *Times,* February 7, 1986,
p. 13.

"Sellafield seeks to end leak fears," Michael Morris and Adrian
Lacey, *The Guardian,* February 7, 1986, p. 1.

"Second strike at Sellafield plant construction site," Adrian Lacey,
The Guardian, February 8, 1986, p. 2.

"Sellafield staff found 'neurotic and unstable,' " Paul Brown, *The
Guardian,* February 11, 1986, p. 32.

"Sellafield workers warned to keep silent," Adrian Lacey, *The Guardian*, February 12, 1986, p. 32.

"New Sellafield scandal: Government admits true level of radiation was concealed for 30 years," *The Sunday Times*, February 16, 1986, p. 1.

"Greenpeace plan to cut radiation," Pearce Wright, *The* (London) *Times*, February 17, 1986, p. 2.

"New nuclear evidence will not change cancer report," Peter Murtagh, *The Guardian*, February 17, 1986, p. 2.

"Dounreay leukemia facts sought," Jean Stead, *The Guardian*, February 18, 1986, p. 4.

"Sellafield: the gloss on the facts," editorial, *The Guardian*, February 18, 1986, p. 10.

"Safety inquiry ordered after Sellafield leaks," John Carvel and Michael Morris, *The Guardian*, February 19, 1986, p. 1.

"Third Sellafield leak fuels closure calls," Michael Morris, *The Guardian*, February 19, 1986, p. 1.

"BNFL denies Greenpeace charge that it lied," Michael Morris and Richard Norton-Taylor, *The Guardian*, February 20, 1986, p. 3.

"Euratom pushes at closed door," Michael Morris and Richard Norton-Taylor, *The Guardian*, February 20, 1986, p. 3.

"Fitzgerald fears dismissed," Anthony Bevins, *The* (London) *Times*, February 20, 1986, p. 1.

"Inquiry ordered into safety at Sellafield plant," Peter Davenport, *The* (London) *Times*, February 20, 1986, p. 1.

"Sellafield: Switch on to the positive," then-BNF chairman Con Allday, *The* (London) *Times*, February 20, 1986, p. 12.

"EEC reject nuclear inspectorate," Derek Brown, *The Guardian*, February 21, 1986, p. 30.

"Radiation tests ordered on Sellafield food," Richard Norton-Taylor, *The Guardian*, February 21, 1986, p. 30.

"Edge of darkness," Geoffrey Lean, *The Observer*, February 23, 1986, p. 11.

"Sellafield: A bad case of exposure?" *The Sunday Times*, February 23, 1986, p. 17.

"Sellafield bosses face shake-up," Geoffrey Lean, *The Observer*, February 23, 1986, p. 1.

"Sellafield: Leaky as a sieve," *The* (London) *Times*, Robin Russell-Jones, February 26, 1986, p. 14.

"Britain to let Europe's inspectors into Sellafield," Steve Connor, *New Scientist*, February 27, 1986, pp. 14–15.

"How safe is Sellafield? Claims about nuclear leaks should be handled with care," James Wilkinson, *The Listener*, February 27, 1986, pp. 2–4.

"Is this the end of the line for Sellafield?" Roger Milne, *New Scientist*, February 27, 1986, pp. 14–15.

"Ministers reveal short list for nuclear dumps," *New Scientist*, February 27, 1986, p. 13.

"Sellafield's safety under the microscope," *New Scientist*, February 27, 1986, pp. 14–15.

"Shut this open sewer," editorial, *New Scientist*, February 27, 1986, p. 12.

"N-plants 'face destruction' in earth tremor," Paul Brown, *The Guardian*, February 28, 1986, p. 1.

"Surviving in the nuclear shadow," Sally Brompton, *The* (London) *Times*, February 28, 1986, p. 15.

"New checks on tremor threat to nuclear reactors," Paul Brown, *The Guardian*, March 1, 1986, p. 1.

"Why shouldn't we be scared when they won't tell us the truth?" Anthony Tucker, *The Guardian*, March 1, 1986, p. 8.

"N-plant hit by fresh leak," Geoffrey Lean, *The Observer*, March 2, 1986, p. 1.

"Minister rejects Sellafield closure," Paul Keel, *The Guardian*, March 3, 1986, p. 30.

"Tories' A-dump disquiet," Alan Travis, *The Guardian*, March 4, 1986, p. 4.

"MPs challenge Sellafield operation," Paul Brown, *The Guardian*, March 6, 1986, p. 1.

"Gas leak 'posed threat for miles,' " Tony Heath, *The Guardian*, March 7, 1986, p. 6.

"I love nuclear power (but not in my backyard)," Rob Edwards, *New Statesman*, March 7, 1986, pp. 10–11.

"Thatcher backs Sellafield waste plant safety record as 'excellent,' " Paul Brown, *The Guardian*, March 8, 1986, p. 32.

"Westminster's nuclear time bomb," Andrew Rawnsley, *The Guardian*, March 12, 1986, p. 21.

"Dumping of waste 'amateurish and haphazard,' " *The Guardian*, March 13, 1986, p. 3.

"MPs challenge case for new nuclear fuel plant," David Fairhall, *The Guardian*, March 13, 1986, p. 3.

"MPs say work on £1.4bn new Sellafield plant should be halted," David Fairhall and Colin Brown, *The Guardian*, March 13, 1986, p. 1.

"Time to call time at Sellafield," editorial, *The Guardian*, March 13, 1986, p. 14.

"Chronicle of wasted cash and kin," David Alton, *The Guardian*, March 14, 1986, p. 16.

"Report backs sea dumping," Colin Brown, *The Guardian*, March 14, 1986, p. 1.

"To lose two tonnes of plutonium is suspiciously sloppy," Keith Barnhan, David Hart, Jenny Nelson, Rob Stevens, *The Guardian*, March 14, 1986, p. 18.

"Sellafield: Profits are not paramount," editorial, *The Observer*, March 16, 1986, p. 8.

"Walker takes up cudgels on behalf of Sellafield," Geoffrey Lean, *The Observer*, March 16, 1986, p. 3.

"BNFL 'did not pass earthquake doubts to NII,' " Paul Brown, *The Guardian*, March 20, 1986, p. 2.

"Civil plutonium was used by military, says CEGB," Richard Norton-Taylor, *The Guardian*, March 21, 1986, p. 2.

"Cold kills 7,000 more than expected," Andrew Veitch, *The Guardian*, March 22, 1986, p. 32.

"Sellafield study 'inadequate,' " Paul Brown, *The Guardian*, March 27, 1986, p. 3.

"Commander accuses State over Hilda Murrell murder," Nick Davies, *The Observer*, March 30, 1986, p. 7.

"Nuclear plant poll fuels concern," Geoffrey Lean, *The Observer*, March 30, 1986, p. 8.

"Island fights mini-Sellafield," Robin McKie and Geoffrey Lean, *The Observer*, April 6, 1986, p. 5.

"A new broom at Sellafield," BNF chairman Christopher Harding, *The Observer*, April 6, 1986, p. 10.

"Dangerous legacy in store for the future," Robin McKie, *The Observer*, April 27, 1986, p. 5.

"The Chernobyl syndrome: The day the impossible happened," Geoffrey Lean, Robin McKie, Andrew Wilson, Peter Pringle, *The Observer*, May 4, 1986, pp. 9–10.

"Nuclear threat put out of harm's way," Roger Highfield, *The Observer*, May 4, 1986, p. 37.

"Clampdown at Sellafield," *New Scientist*, May 8, 1986, p. 22.

"Nuclear propaganda," editorial, *New Scientist*, May 8, 1986, p. 16.

"Confusion in Britain: Call up the rubber boots," Paul Lashmar, Robin McKie, Geoffrey Lean, *The Observer*, May 11, 1986, p. 11.

"A lot of fuss about a few millisieverts," Sharon Kingman, *New Scientist*, May 15, 1986, p. 26.

"MPs 'created the nuclear mess that they condemn,' " *New Scientist*, May 15, 1986, p. 26.

"Massive Nuclear Site Disturbs Britons," Karen DeYoung, *The Washington Post*, May 19, 1986, p. Al.

"In England, a Nuclear Plant Slowly Poisons Land and Sea," Patrick J. Sloyan, New York *Newsday*, May 20, 1986, p. 1.

"Introducing the national cabbage monitoring network," Ian Mason, *New Scientist*, May 22, 1986, p. 23.

"The nuclear watchdog strains at the leash," Michael Kenward, *New Scientist*, May 22, 1986, pp. 58–9.

"Radiation monitors hit cash crisis," *New Scientist*, May 22, 1986, p. 23.

"Sellafield spurns the bunker image," Colin Smith, *The Observer*, May 25, 1986, p. 52.

"Sellafield in 'lost' nuclear fuel blunder," John Sweeney, *The Observer*, June 8, 1986, p. 1.

"Ackworth takes charge," Grenville Needham, *New Scientist*, June 12, 1986, pp. 57–8.

"Doctors on panels told to oppose secrecy," Pearce Wright, *The (London) Times*, June 13, 1986, p. 3.

"Radioactive meat was on sale for a month," Geoffrey Lean, *The Observer*, June 22, 1986, p. 1.

"MP's ire at meat 'secrecy,' " *The Observer*, June 29, 1986, p. 5.

"Nuclear family haunted by a testing legacy," Geoffrey Lean, *The Observer*, June 29, 1986, p. 55.

"The Saga of Windscale: Profits At Any Cost," Andre Carothers, Greenpeace *Examiner*, June 1986, p. 20.

"Your daily dose of radiation," Geoffrey Lean, *The Observer*, July 13, 1986, p. 49.

"Sellafield workers gagged by the rule book," Geoffrey Lean, *The Observer*, August 24, 1986, p. 5.

"Experts to vet British reactor," Geoffrey Lean, *The Observer*, August 31, 1986, p. 1.

"Majority say No to nuclear power," Steve Vines, *The Observer*, September 14, 1986, p. 1.

"The case for nuclear power in Britain," Alan Cottrell, *The Observer*, September 28, 1986, p. 12.

"Sellafield under suspended sentence," Roger Milne and Fred Pearce, *New Scientist*, December 18, 1986, pp. 10–11.

"Britain watered down anti-pollution rules," Geoffrey Lean, *The Observer*, July 5, 1987, p. 3.

"Seeping threat to Seal Sands swans," Victor Smart, *The Observer*, August 2, 1987, p. 5.

"Unions try to stem rising tide of toxins," Victor Smart, *The Observer*, August 2, 1987, p. 5.

"UKAEA Report Indicts Soviet RBMK Design," *Nuclear News*, August 1987, pp. 66, 68.

"NRPB Study Dismisses Child Leukemia Link in U.K.," *Nuclear News*, September 1987, pp. 107–8.

"Shock plan on atom disasters," Tony Heath and Geoffrey Lean, *The Observer*, October 4, 1987, p. 1.

"Japan Increasing Its Nuclear Power," Walter Sullivan, *The New York Times*, October 9, 1987, p. 9.

"Atomic mop brigade," David Siddall and Geoffrey Lean, *The Observer*, October 11, 1987, p. 7.

"Britain dodges ban on Botha's uranium," Martin Bailey, *The Observer*, October 11, 1987, p. 7.

"Rising Nuclear Trade Stirs Fear of Terrorism," John H. Cushman, Jr., *The New York Times*, November 5, 1987, p. 5.

"Britain set to defy North Sea dump plea," Geoffrey Lean, *The Observer*, November 22, 1987, p. 5.

"Mock invasion sent to nuclear hot-spot," Ian Mather, *The Observer*, November 22, 1987, p. 3.

"Poison dump to be sold for housing," Geoffrey Lean, *The Observer*, November 29, 1987, p. 5.

"Bunkers built for N-waste 'mountain,' " Paul Lashmar and Andrew Cavenagh, *The Observer*, October 16, 1988, p. 4.

Nuclear Bibliography

Anonymous. *Nuclear Energy in Britain*. London: Her Majesty's Stationery Office, 1976.

——. *Nuclear Energy in Britain*. London: Her Majesty's Stationery Office, 1981.

——. "The Windscale File: A Lay-Guide to Living (and Dying) with a Nuclear Neighbor." London: Greenpeace Limited, 1984.

Bunyard, Peter. *Nuclear Britain*. London: New English Library, 1981.

Cutler, James, and Rob Edwards. *Britain's Nuclear Nightmare*. London: Sphere Books Limited, 1988.

Goldsmith, Edward, and Nicholas Hildyard, eds. *Green Britain or Industrial Wasteland?* Cambridge, England: Polity Press, 1986.

Grover, J. R., ed. *Management of Plutonium Contaminated Waste*. London: Harwood Academic Publishers, for the Commission of the European Communities, 1981.

Hawkes, Nigel, and Geoffrey Lean, David Leigh, Robin McKie, Peter Pringle, Andrew Wilson. *Chernobyl: The End of the Nuclear Dream*. New York: Vintage Books, 1987.

Meyer, Stephen M. *The Dynamics of Nuclear Proliferation*. University of Chicago Press, 1984.

Patterson, Walter C. *The Plutonium Business and the Spread of the Bomb*. San Francisco: Sierra Club Books, for the Nuclear Control Institute, 1984.

Roberts, Alan, and Zhores Medvedev. *Hazards of Nuclear Power*. Nottingham: Russell Press Ltd., 1977.

Schell, Jonathan. *The Fate of the Earth*. New York: Avon, 1982.

Smith, Joan. *Clouds of Deceit: The Deadly Legacy of Britain's Bomb Tests*. London: Faber and Faber, 1985.

Social Bibliography

Addams, Jane, Bernard Bossanquet et al. *Philanthropy and Social Progress*. Boston: Thomas Y. Crowell and Co., 1893.

Ashford, Douglas E. *Policy and Politics in Britain: The Limits of Consensus*. Philadelphia: Temple University Press, 1981.

——. *The Emergence of the Welfare States*. Oxford: Basil Blackwell, 1986.

Belloc, Hilaire. *The Servile State*. London: T. N. Foulis, 1912.

Benn, Tony. *Parliament, People and Power: Agenda for a Free Society*. London: Verso, 1982.

Bentham, Jeremy. *Panopticon; or the Inspection House*, in *The Works of Jeremy Bentham*, Vol. IV. Edinburgh: William Tait, 1843.

Beveridge, Janet. *Beveridge and His Plan*. London: Hodder and Stoughton, 1954

Beveridge, William H. *Social Insurance and Allied Services*. New York: Macmillan, 1942.

——. *Unemployment: A Problem of Industry*. London: Longmans, Green and Co., 1930

——. *Voluntary Action: A Report on Methods of Social Advance*. London: George Allen and Unwin, 1948.

Blackwell, Trevor, and Jeremy Seabrook. *A World Still to Win: The Reconstruction of the Post-War Working Class*. London: Faber and Faber, 1985.

Bland, A. E., P. A. Brown, R. H. Tawney. *English Economic History: Select Documents*. London: G. Bell and Sons, 1914.

Booth, Charles. *Pauperism and the Endowment of Old Age*. London: Macmillan and Co., 1892.

Byllesby, L. *Observations on the Sources and Effects of Unequal Wealth; with propositions towards Remedying the Disparity of Profit in pursuing the Arts of Life and Establishing Security in Individual prospects and resources*. New York: Lewis J. Nichols, 1826.

Chalmers, Thomas. *On Political Economy in Connexion with the Moral State and Moral Prospects of Society*. Glasgow: William Collins, 1832.

Chesterton, G. K. *Eugenics and Other Evils*. New York: Dodd, 1927.

Dahrendorf, Ralf. *On Britain*. University of Chicago Press, 1982.

Defoe, Daniel. *Giving Alms, no Charity and Employing the Poor a grievance to the nation*. London: 1704; Yorkshire: S. R. Publishers, 1970.

Eden, Frederick Morton. *The State of the Poor*, ed. A.G.L. Rogers. London: G. Routledge & Sons, 1928 [c. 1797].

Fielding, Henry. *An Enquiry into the Causes of the Late Increase of Robbers and Related Writings*. Oxford: Clarendon Press, 1988.

Fried, Albert, and Richard Elman, eds. *Charles Booth's London*. New York: Random House, 1968.

George, Henry. *Progress and Poverty*. New York: Doubleday, Page, 1903 [c. 1879].

Greeley, Horace. *The American Laborer*. New York: Greeley and McElrath, 1843.

———. *Essays Designed to Elucidate the Science of Political Economy*. Boston: Fields, Osgood and Co., 1870.

Havighurst, Alfred F. *Britain in Transition*. University of Chicago Press, 1985.

Hill, Christopher. *Intellectual Origins of the English Revolution*. Oxford: Clarendon Press, 1982.

Himmelfarb, Gertrude. *The Idea of Poverty: England in the Early Industrial Age*. New York: Vintage Books, 1985.

Hobsbawm, E. J. *Industry and Empire*. London: Penguin Books, 1984.

Jarman, T. L. *Socialism in Britain*. New York: Taplinger Publishing Company, 1972.

Jones, Catherine, and June Stevenson, eds. *The Year Book of Social Policy in Britain 1982*. London: Routledge and Kegan Paul, 1983.

———. *The Year Book of Social Policy in Britain 1983*. London: Routledge and Kegan Paul, 1984.

Jones, Gareth Stedman. *Outcast London*. New York: Pantheon Books, 1984.

Kirkup, Thomas. *A History of Socialism*. London: Adam and Charles Black, 1906.

MacKillop, Ian. *The British Ethical Societies*. Cambridge University Press, 1986.

Malthus, Thomas. *An Essay on the Principle of Population*. ed. Anthony Flew, 1970.

Marx, Karl. *Capital.* New York: Modern Library, 1906.

Owen, Robert. *A New View of Society.* London: J. M. Dent & Sons, 1927.

Perkin, Harold. *Origins of Modern English Society.* London: Ark Paperbacks, 1985.

Pigou, A. C. *Wealth and Welfare.* London: Macmillan and Co., 1912.

Rose, Richard. *Politics in England.* London: Faber and Faber, 1985.

Shaw, G. Bernard, ed. *Fabian Essays in Socialism.* New York: Dolphin Books, n.d.

Sked, Alan, and Chris Cook. *Post-War Britain.* London: Penguin Books, 1984.

Smith, Adam. *An Inquiry into the Nature and Causes of the Wealth of Nations.* University of Chicago Press, 1976.

Stevenson, John. *British Society 1914–45.* Harmondsworth: Penguin Books, 1984.

Stewart, Angus. *Contemporary Britain.* London: Routledge and Kegan Paul, 1983.

Stowe, Harriet Beecher. *The Lives and Deeds of Our Self-Made Men.* Boston: Estes and Lauriat, 1889.

Thompson, Dorothy. *The Chartists: Popular Politics in the Industrial Revolution.* New York: Pantheon Books, 1984.

Thompson, E. P. *The Making of the English Working Class.* New York: Vintage Books, 1966.

Townsend, Joseph. *A Dissertation on the Poor Laws.* Berkeley and Los Angeles: University of California Press, 1971.

Waller, P. J. *Town, City and Nation: England 1850–1914.* Oxford University Press, 1983.

Webb, Beatrice. *"Break Up the Poor Law and Abolish the Workhouse": Being Part I of the Minority Report of the Poor Law Commission.* London: The Fabian Society, 1909.

Weightman, Gavin, and Steve Humphries. *The Making of Modern London: 1815–1914.* London: Sidgwick and Jackson, 1983.

Wells, H. G. *Tono-Bungay.* New York: Modern Library, 1935.

Wilding, Paul, ed. *In Defense of the Welfare State.* Manchester University Press, 1986.